LAND ROVER 90 AND 110
OWNER'S AND BUYER'S GUIDE

by James L. Taylor

1995

Published and distributed by:
YESTERYEAR BOOKS
60 Woodville Road
London NW11 9TN
(081-455 6992)

ISBN 1-873078-17-X

CONTENTS

Introdution 3

The Land Rover Marque 4

Development 5

Configuration 9

The First One Tens, 1983 12

Buying an Early 110 16

The Crew Cab and the 2.5 Diesel, 1984 18

Buying a 1984 Model 20

The 90 and the 1985-Model 110 21

The V8 90 and 2.5 Petrol Engine For 1986 23

Buying a 1985 or 1986 Model 90 or 110 24

1987 Models: The Diesel Turbo and Uprated V8 25

Buying a 1987 Model 28

The Facelift for 1988 29

Buying a 1988 Model 33

The 1989-1990 Season Models 34

Buying A 1989 or 1990 Model 36

Enter the Defender for 1991 37

Military 90s, 110s and 127s 37

Australian Peculiarities 42

Factory-Approved Conversions 45

Modifications 51

Buying a Converted or Modified Vehicle 52

90s and 110s in the Camel Trophy 53

Key Dates 55

Miscellany 56

Specifications (Home Market Models) 60

Vehicle Identification 61

Colour Chart 62

Road Performance Figures 63

Production Figures 63

INTRODUCTION

Even though the 90 and 110 family has only recently gone out of production, its members already arouse great interest among enthusiasts. So much so, in fact, that the mail-order bookshop of Land Rover Owner magazine approached Yesteryear Books during 1993 and practically begged for a book on the subject which they could sell. Here, then, is that book.

The sheer variety of models which Land Rover Ltd produced within their 90 and 110 range during the 1980s is astonishing - and not a little bewildering. Even those who work with the vehicles are often confused by specification changes, and those who are simply in the market to buy one can rapidly end up with information, both correct and misleading, flowing out of their ears. So one of the aims of this book is to provide a core of reliable and useful information, for both enthusiasts and those who simply want to buy a used Land Rover.

Land Rovers are rarely totally "standard" because they are essentially practical vehicles which owners often modify in use to suit their individual requirements. For this reason, it can be helpful when buying a vehicle to know what is "standard" and what is not, if only because the factory's Workshop Manual and Parts Books will not provide assistance on non-standard items. Far better to recognise from the outset how much of a problem this is likely to be than to find out the hard way when the vehicle breaks down! This book therefore aims to establish "standard" specifications as far as possible, to help both those who aim to keep a vehicle in original condition and those who need to establish exactly what has been modified on a vehicle.

The book's format is chronological. It traces the development of the coil-sprung Land Rovers between 1983 and 1990, when the original range was replaced by Defender derivatives. Each stage in the vehicles' development is discussed in detail, with specification changes and other useful points. After each of the descriptive sections comes a detailed guide on the strengths and weaknesses of the models in question. The advice this contains is designed not only to assist buyers in their choice of model, but also to warn them about potentially expensive faults. However, because major items like engines or gearboxes are dealt with only once (on their first introduction), it is important to read the whole book and not just the section relating to a model of particular interest.

Very many Land Rovers formed the basis of specialist conversions when they were new, and there is a section in this book devoted to them. Equally, many DIY owners have modified vehicles to suit their requirements, and there is a section on such modifications. Finally, the book includes some helpful specification tables and statistical data which will be of interest to the Land Rover owner.

Although neither the author nor the publisher can accept any liablility for acts carried out on the advice given in this book, both would be pleased to hear about any errors in it or omissions from it.

December 1993   James Taylor

THE LAND ROVER MARQUE

Even though the company which made the 90 and 110 models was not formed until 1978, the Land Rover marque was actually born some 30 years before that. After the Second World War ended in 1945, the Rover Company had some difficulty in adapting to post-War trading conditions, as its high-quality luxury cars of essentially pre-War design did not sell well in the overseas markets which the British Government insisted should be the car manufacturers' highest priority. In order to improve their overseas sales performance, Rover therefore developed a small and rugged utitlity vehicle which was intended to suit farmers and others who worked on the land. This vehicle was aptly christened the Land-Rover.

The Land-Rover rapidly became a runaway success, and Rover abandoned their original plan of taking it out of production as soon as normal trading conditions returned to the luxury car market. For the next three decades, Land-Rovers sold in far greater numbers than Rover cars, and there is no doubt that it was profits from Land-Rover sales which kept the company afloat in the 1950s and 1960s, even though Rover always saw themselves primarily as car makers.

Changes elsewhere in the British car industry forced Rover to join up with the Leyland truck and bus combine in the mid-1960s, and from 1968, they came into the British Leyland combine when Leyland merged with British Motor Holdings. Unfortunately, the British Leyland story was not a happy one. The company was forced to put most of its profits (of which a substantial proportion came from Land-Rovers) into its loss making Austin Morris Volume Cars Division and Land-Rovers were granted very little development capital for the design and production of new vehicles. Although the Range Rover of 1970 proved another runaway success, Land-Rovers were obliged to soldier on through the 1970s with an outdated design.

The inevitable process of decline had already started when British Leyland decided to invest heavily in the future of Land-Rovers. In 1978, the company established Land-Rover Ltd (the hyphen disappeared from the name and from the vehicles some two years later), and the modern success story of the marque began. Production was expanded, existing models were improved, and eventually new models - the 90 and 110 among them - were announced. Their arrival coincided with a growing international interest in four-wheel-drive passenger vehicles, and the Land Rover 90 and 110 remained among the most respected workhorse vehicles throughout their production run between 1983 and 1990.

DEVELOPMENT

Among the earliest prototypes were some built on the Range Rover's 100-inch wheelbase chassis. This one, registered as CRW 453T probably in 1978, seems to have been used for styling trials. The black wheelarch eyebrows, single-piece windscreen and shaped bonnet are already in place, but the rear of the cab and the grille differ from production versions, and the styling crease on the side of the pick-up body does not mate up happily with the line on the front wings and the doors. The black bumpers did not appear when production of the coil-sprung Land Rovers began in 1983.

The late 1960s and early 1970s were a confusing and demoralising period for those involved with the design, manufacture and sales of Land Rovers. As British Leyland - formed in 1968 - gradually tried to bring some order to the vast motor manufacturing empire it had inherited, money in some areas became very tight. At the Solihull home of the Land Rover, 1970's triumphant launch of the Range Rover was followed by the disappointing 1971 launch of the Series III Land Rover, which was little more than a warmed over version of the existing models because there had not been enough money to develop another new vehicle.

The Series III continued to sell strongly simply because it was a Land Rover, but the more perceptive designers

Despite its 1983 registration number, this 90 prototype - the first one to be made - dates from some years earlier. The one-piece windscreen was made to look like a Series III two-piece type with a strip of black tape, and the grille panel comes from a Series III V8 W9, perhaps again in order to disguise the vehicle when it was out on road test. This vehicle is now in the Dunsfold Land Rover Trust collection.

and engineers at Solihull knew that it was rapidly becoming an anachronism and would need to be replaced before long. The British Leyland corporate view, however, was that the vehicle was selling well enough to need no new investment of capital for the time being, and the company therefore devoted the major proportion of the development funds it had to other areas.

Very little happened during the early 1970s, but the injection of Government money in 1975 to keep the company afloat seems to have been the catalyst which spurred the Land Rover designers into action again. Under the Ryder plan which was supposed to put British Leyland back in the black, £400 million was proposed to double Land Rover and Range Rover production. Anticipating that this massive injection of capital would eventually allow for the introduction of new models, Land Rover Director of Engineering Tom Barton asked his designers to start looking at what these new models might be like.

However, there was no money available just yet, and so Barton's engineers began by drawing on existing production components as much as they could. The main failings of the Series III Land Rovers then in production

were that they were unrefined and offered poor road performance, but the Land Rover engineers had already successfully tackled both of these problems for the highly-acclaimed Range Rover by using coil springs instead of leaf springs and by fitting the Rover V8 petrol engine. The obvious way to go was therefore to put a Land Rover body onto a Range Rover chassis and drivetrain, and that was exactly what the engineers did. The resulting hybrid was very crude in some respects - the long-wheelbase Series III Land Rover body had to be shortened by nine inches in the wheelbase and by several inches at the rear in order to fit the Range Rover chassis - but it was effective enough to persuade Tom Barton that the combination had merit.

A single wheelbase was not going to be enough, however. Since 1953, an important element in the Land Rover's success had been that it was available with a short wheelbase for runabout duties or a long one for more demanding load-carrying work. So the designers started to look at ways of developing the Range Rover chassis with its 100-inch wheelbase into a pair of chassis which would more closely match the existing 88-inch and 109-

Final styling details were not settled until quite late on. This 110 High Capacity Pick-up pictured in July 1982 has an Ivory rear body as well as cab roof, and black wheel centres and bulkhead air vents. The extended grille panel was used on air conditioned vehicles. Note the starting-handle bracket, which was not present on production models. The 1980-1981 number plate almost certainly did not belong to the vehicle but was simply added for the photograph.

inch wheelbases of the Series III Land Rovers.

In fact, the Range Rover chassis was already available for special bodywork with a 110-inch wheelbase through converters Spencer Abbott, and it may have been this happy coincidence which led to the adoption of a 110-inch wheelbase for the new long-wheelbase Land Rover. As for the short-wheelbase model, the Range Rover chassis was shortened to give a 90-inch wheelbase, and a V8-powered 90-inch Land Rover existed in prototype form as early as January 1977. This early prototype was designed specifically for military use, probably with the intention that it should eventually replace the 88-inch Military Lightweight Land Rovers which had gone out of production in 1975.

During 1977, the project gathered momentum. A design team was appointed consisting of Chief Engineer Mike Broadhead with Bob Lees and Brian Anderson. And over the next few years, a whole series of prototypes appeared, most of them cobbled together using existing production parts wherever possible. However, an important turning point came in 1978. Plans to separate Land Rover and Range Rover production from car

production within British Leyland had been brewing for some time, and in July that year Land Rover Ltd was created as a new division within British Leyland. A month later, the company announced that it had been granted £280 million to ensure its long-term survival, and that this money would fund a more limited expansion of production than had been proposed under the 1975 plan and, more important, that it would also fund several new models.

The funds were allocated in stages, and the first stage was implemented almost immediately. Early in 1979, a V8-engined Series III 109-inch Land Rover became available, giving a much needed performance boost to the old warhorse. Next came a series of important specification improvements for the Range Rover and a major reorganisation of production facilities at Solihull. Under this scheme, the Range Rover assembly lines were moved into the former P6 car assembly plant (which had been empty since production had ended in 1976) and space was freed up in the old South Works for an assembly line to build the new generation of Land Rovers.

While all this was going on, prototypes of the new

That same number plate turned up again on this styling vehicle, a 110 Station Wagon pictured on the same day as the pick-up. The principle of the body-coloured wheelarch eyebrows had been established by this stage, and the Station Wagons would go into production with those Ivory-coloured upper panels. The black wheel centres were not approved for production, however, and there would be a special grille badge in place of the Stage I V8 item seen here.

models were completing more than a million miles of testing both at home and abroad. The 90-inch chassis of the short wheelbase model had been stretched to 92.9 inches to bring it closer in size to competitors from Mitsubishi, Nissan, Jeep and others, and the 110-inch chassis had been strengthened to cope with the demands users were expected to make of it. Styling details had also been settled. As a decision had been taken to delay the introduction of the short-wheelbase variant by a year or so and to introduce the long-wheelbase models first, styling details were developed primarily on the 110-inch models. Most of the work was done during 1981 and 1982, under the direction of Tony Poole.

By the end of 1982, everything was in place, and the first Land Rover 110s came off the assembly lines early in the new year. For the time being, the Series III 88 inch and 109 inch models would remain in production, but in due course they were to be superseded by 90s and 110s.

C O N F I G U R A T I O N

Common to all the coil-sprung Land Rovers was this front suspension layout, with disc brakes.

Certain basic design elements are common to all the Land Rovers covered in this book. Those elements were incorporated into the two models which were central to the Land Rover range of the 1980s - the 90 and 110 - and it is those two models on which other variants were based. However, even though the 127 and the Australian 120 inch and 6 x 6 models shared several major elements of this same basic design, each one also had its own special differences. These are discussed elsewhere. While the 90 and 110 were in production, Land Rover Santana in Spain was also building Land Rovers - but these were developments of the older Series III models and are not covered here.

The 90 and 110 models derive their strength from a rigid steel ladder-frame chassis, built up from box-section side members and cross-members. Some of these cross-members are bolted to the main frame so that they can be removed to give access to transmission components. All elements of the chassis frame are precision-welded and electrophoretically corrosion-proofed but, contrary to popular belief, they are not galvanised. Jacking points are incorporated into the chassis frame at the front and rear of each side-member. Although the general design of the 90 and 110 frames is similar, the side-members of the long-wheelbase type are deeper, to provide additional strength.

Earlier Land Rovers had semi-elliptic springs, but the suspension on the 90 and 110 has long-travel coil springs, together with telescopic dampers. Like all Land Rovers since 1948, however, the 90 and 110 have the beam axles

Coil-sprung rear suspension - with drum brakes - was fitted to all models except the Australian - built 6 x 6.

which Land Rover designers have always insisted give better off-road performance than independently sprung wheels. At the front, the axle is located by radius arms and a Panhard rod ; at the rear, there are radius arms and support rods for fore-and-aft location, with a central A-frame for sideways location. On some models, this A-frame incorporates a Boge self-energising ride-levelling strut, which maintains constant ride height regardless of the load in the vehicle and thus allows the coil springs to retain their full travel at all times.

All the 90 and 110 family of Land Rovers have recirculating-ball, worm-and-nut steering, of which a power-assisted version was standard on some models and optional on others. A transverse steering damper on the front axle prevents severe kickback at the steering wheel when the vehicle is being driven in rough terrain. There are disc brakes on the front wheels and drum brakes at the rear - a system which gives much better stopping power than the all-drum system on earlier Land Rovers. The hydraulic braking circuit has dual lines for safety and has vacuum servo assistance. Unlike a conventional car, the parking brake does not operate on the wheels but is contained in a separate drum which acts on the transmission output shaft.

Like all Land Rovers, the members of the 90 and 110 family have a dual range transmission. The conventional four-speed or five-speed manual gearbox drives the wheels through a two-speed transfer box. The "high" range gear of this transfer box is designed for road use, while the "low" range multiplies engine torque for increased traction on difficult terrain. The transfer gears operate on reverse as well as forward gears. Although some versions of the 110 were initially available with the same selectable two-wheel drive as had characterised all earlier Land Rovers, the alternative permanent four-wheel-drive proved far more popular, and was standardised early on. With permanent four-wheel-drive, there is a third differential in the transmission, which prevents the transmission from being damaged when (as regularly happens) front and rear axles rotate at slightly different speeds. This third or centre differential can be locked by the driver to give maximum traction in soft or slippery conditions.

The engines used in the 90 and 110 models were all developments of those used in the Series III Land Rovers, although each was developed further during the 1980s. The four-cylinder petrol and four-cylinder diesel engines were derived from 1950s designs ; initially they displaced 2 1/4 litres, but both were later uprated to 2 1/2 litres. The diesel engine was also redeveloped to take a turbocharger. Top of the range in most markets was the 3 1/2-litre V8 petrol engine, which had its origins in an American General Motors design to which Rover had bought the

All the coil-sprung Land Rovers had a robust ladder-frame chassis.

manufacturing rights in 1964. In Australia, the only diesel model available had a 3.9-litre Isuzu truck engine, seen earlier in Australian-built Series III Land Rovers and later available also with a turbocharger.

In order to retain the Land Rover family identity, the styling of the 90 and 110 bodies was deliberately made to resemble that of the Series III models very closely, although there were in fact no common panels between the older and newer types. These bodies were bolted to the chassis in the same way as they had been on earlier models. They were constructed with corrosion-resistant aluminium outer panels bolted or riveted together, with some steel inner panels for additional strength. Although there were detail differences in the construction, this was the way in which Land Rover bodies had been built ever since 1948.

Some members of the 90 and 110 family were supplied in chassis/cab form, and most of these retained the characteristic Land Rover front end styling. However, the construction of the bodies varied. There were also special widened cabs for the military 6x6 Land Rovers built in Australia, although these followed the general outlines of the standard variety and were instantly recognisable as Land Rover products.

THE FIRST ONE TENS, 1983

Land Rover's new Land Rover, as publicity material called the 110, was introduced at the Geneva Motor Show in March 1983. At first, it was to be available only on the home market and in Switzerland, although it was introduced to other markets gradually over the next few months. According to the plate badge above its radiator grille, it was a Land Rover 110, although most publicity referred to it as a One Ten, simply to avoid the less snappy version of "Hundred and Ten". There had of course been two previous Rover products with similar names: the P4 110 saloon built between 1962 and 1964, and the Series IIB Forward-Control 110-inch Land Rover built between 1966 and 1972.

In most markets (although not at home), the 110 was introduced alongside the Series III 109-inch models, which would remain in production until 1985: according to Land Rover Ltd, this was because many overseas fleet buyers tested a vehicle for as long as three years before placing a bulk order, and in the mean time it was important to continue production of a vehicle which they had already tried and tested. The short-wheelbase 88-inch Series III models remained available in all markets, and Land Rover refused to be drawn on whether there would eventually be a short-wheelbase equivalent of the 110 to replace them. However, the company did admit that the new vehicle had been introduced as a long-wheelbase model because 70% of Land Rover sales worldwide were of long-wheelbase versions; that contrasted with the position in Europe and on the home market, where the short-wheelbase models were always much more popular.

Even though there was a lot that was new about the 110, there was also a lot that was comfortingly familiar. Not only was the general outline of the vehicle very similar to that of the Series III 109, but the familiar range of five body types remained available. These were the soft-top, hardtop, pick-up, High-Capacity Pick-Up and Station Wagon, and the 110 could also be supplied in chassis/cab form for special bodywork. Even so, a 110 was instantly distinguishable from earlier Land Rovers at a glance.

The major differences lay at the front of the vehicle, where there was a single-piece windscreen in place of the divided screen fitted to all of Solihull's earlier Land Rovers. This screen was also taller than before, and offered a 25% increase in area as compared to the Series III type. There was also a new flush front, with a black plastic slatted grille and matching black plastic headlamp surround panels. The heater air intake was now on top of the left-hand wing instead of in its side, and was matched by a dummy intake on the right. And there was a neatly shaped bonnet panel, with a shallow depression in its centre.

Most obvious from the sides were the deformable "eyebrows" fitted over the wheelarches, which were designed to afford a degree of protection to the body sides and to compensate for the widened tracks of the 110. There was also no doubt that they gave the vehicle a rather more aggressive appearance, which must have been considered

an important sales point at a time when the market was receptive to this kind of styling. At the rear, the tail lamps were mounted on the lower body corners instead of at waist level as on the Series III. Station Wagons benefitted from a third hinge on the rear door and from a higher mounting position for the exterior spare wheel : on earlier Station Wagons, the low position of the wheel had made it impossible to fit a tow hitch to the rear cross-member. And lastly, there were three paint colours available for the 110 which had not been offered on Series IIIs: Roan Brown, Stratos Blue and Trident Green.

The Series III Land Rovers had been very spartan inside, and the 110 was almost luxurious by comparison, even though it retained its workmanlike character. The dashboard was a completely new moulding, and the instrument panel ahead of the driver was very much neater and more legible than the Series III type. A four-spoke steering wheel, similar to that on the RangeRover but with a Land Rover logo on its centre pad, replaced the old three-spoke type. Front seats were adjustable for the first time in a Land Rover, and the gear levers and handbrake had been given a neater appearance, which was enhanced by the presence of tough rubber gaiters. Altogether, the interior of the 110 seemed very much more car-like than that of previous Land Rovers.

Road behaviour was also more car-like. Earlier Land Rovers with their rugged leaf spring suspension had given a rough, jolty ride, but the coilsprung 110s rode much like a Range Rover. The ride was altogether softer and more comfortable, although the long-travel springs did allow rather more body roll on corners than the leaf springs had done. In rough terrain, these springs also made the 110 much more capable than its predecessors, because the increased travel (up by 50% at the front and by 25% at the rear) allowed greater axle articulation, which in turn increased the chances that the wheels would remain in contact with the ground in extreme conditions. Braking - which could cause some anxious moments in a heavily-laden Series III - was also vastly improved by the addition of discs at the front and a vacuum servo.

The 110 came with a choice of three engines, all of them familiar from the Series III Land Rovers. Least powerful was the 2,286cc four-cylinder indirect-injection diesel, with just 60 bhp at 4,000 rpm and 103 lb/ft of torque at 1,800rpm. This was an elderly engine, last improved (with a five-bearing crankshaft) in 1980, but dating from 1961 in its original three-bearing form and in turn derived from the very first 2-litre Land Rover diesel of 1957. However, it had always proved popular with fleet operators and with those who used the Land Rover's power take-offs, and for this reason was retained as the rather poor cousin of the two petrol engines. In the 110, it was fitted for the first time with a key-operated (solenoid) cut-out.

The smaller of the two petrol engines had the same

The grille badge on the new models left no doubt about what they were called. The bonnet lock shows that this 1983 vehicle is a County Station Wagon.

The expected range of utility versions was made available from the beginning. This 1983 One Ten Hard Top shows off the production front end styling and the wheelarch eyebrows, which were always painted to match the body at this stage.

The One Ten County Station Wagons were readily distinguishable by their Ivory-coloured side stripes.

2,286cc capacity and the same basic layout as the diesel. It had first been seen in three-bearing form in 1958 and, like the diesel, had been re-engineered in 1980 to take a five-bearing crankshaft. However, for the 110 it had been further improved, with redesigned inlet and exhaust manifolds, a Weber 32/34 DMTL carburettor in place of the earlier Solex, and a revised camshaft. The main aim of the modifications had been to improve low-speed torque, but there had been power gains as well. In the 110, what Land Rover now called the 2.3-litre petrol engine (to distinguish it from the older "2 ¼" version) had 74 bhp at 4,000 rpm and 163 lb/ft at 2,000 rpm.

Top of the engine range was the 3 1/2 litre V8 petrol engine, uprated by 25% as compared to the version which had been offered in the Series III ("Stage I") 109 inch models. Power was increased to 114bhp at 4,000 rpm and torque to 185 lb/ft at 2,500 rpm. This was the engine which would prove most popular in the passenger-carrying versions of the 110, because it offered refinement, great flexibility, and the ability to cruise at 80mph or more on motorways.

The only transmission available with the V8 engine was a four-speed type with permanent four-wheel drive. However, the two four-cylinder models came with five-speed transmissions, and could be ordered with permanent four-wheel drive or with the selectable four-wheel drive used in earlier Land Rovers. In all cases, the main gearboxes drove through two-speed transfer gearboxes.

The four-speed gearbox in the V8 models was the LT95 type used in the Range Rover, which came with an integral transfer box. The four-cylinder 110s had the LT77 five-speed, originally developed for Rover saloons in the mid-1970s and also seen in Triumph sports cars and Jaguar saloons. This came with a separate transfer box, the LT230R type which had first been seen in the automatic Range Rover introduced in August 1982.

The 110 was an immediate success. Despite the late start

to sales, 110 models accounted for almost a third of home market Land Rover sales during 1983. It was clearly taking some sales away from the short-wheelbase Series IIIs as well as effectively replacing the 109-inch models, and the Station Wagons began to exert a strong appeal outside the traditional Land Rover territory. This was partly because some buyers saw them as a cut-price Range Rover alternative, but also because their higher refinement levels made them for the first time a serious alternative to conventional two-wheel-drive estate cars.

The biggest success belonged to the top-model County Station Wagons, however. Sales exceeded Land Rover's own expectations, and it was these vehicles which really created the 110's image in the home market and, later, in other developed countries. The County Station Wagons came with a great deal of special equipment, some of which was also optionally available on other models, and could be instantly recognised by their side stripe decals and "County" badges on the sides and rear. Special equipment included tinted glass, brown cloth upholstery, rubber pedal pads, halogen headlamps, front and rear mud flaps, a reversing lamp, a dipping interior mirror, a clock, a volt meter, an ash tray and sun visors. The headlining was made of resin-impregnated felt, which had good insulating properties and so made it unnecessary to fit the Tropical roof of earlier Station Wagons. There were side repeater flashers on the front wings, a hazard warning system, a spare wheel cover and a bonnet lock. In addition, there were self-levelling rear suspension, power-assisted steering and radial tyres in place of the crossplies fitted to other models.

As Land Rover customers expected, a wide range of optional equipment was available. This covered both utility equipment, such as winches and power take-offs, and other types of extras. These were available either individually or grouped into Option Packs covering Electrical, Protection, Towing and Interior Appointment additions.

The dashboard was very much neater and better-finished than that on earlier Land Rovers. This is a One Ten County Station Wagon, complete with headrests on the outer pair of front seats. The markings on the gear lever knob show that a five-speed gearbox is fitted, so this vehicle must have one of the two four-cylinder engines.

Door trims on the County, while hardly the last word in early 1980s sophistication, were very much more car-like than anything ever fitted to earlier Land Rovers.

BUYING AN EARLY 110

When the 110 was announced in March 1983, Y-suffix registrations were current in Great Britain. The earliest examples, including the factory press demonstrators, therefore all have these registrations. Chassis numbers do not give any indication of build date (although a comparison with the lists at the back of this book will give an approximate date), and therefore a vehicle's registration number is likely to provide the first evidence that a vehicle is one of these early 110s.

All these first 110s were very much cruder than their later counterparts. Soundproofing was rudimentary ; interior trim was spartan ; the windows slid open in channels (which made them easier for thieves to open) ; and road performance was relatively poor. For all these reasons, they may not be the ideal choice for many buyers. Nevertheless, they are rugged and dependable vehicles, and buyers who are attracted to a 110 for its abilities as a workhorse need not be discouraged by these shortcomings. Not surprisingly, prices of the earliest vehicles are also generally lower than those asked for later examples.

Obviously, indications such as mileometer reading and general condition will help to determine whether an early 110 is a sensible purchase. However, there are also several less obvious points to watch for when buying. Many people, for example, firmly believe that all Land Rovers have galvanised chassis and that these chassis therefore never rust. In fact, only the first pre-production models built in 1948 had galvanised chassis, and the chassis of an early 110 can and does rust.

Even though the 110 chassis is an immensely strong box-section affair and will therefore resist corrosion for quite a long time, it is always advisable to check thoroughly for signs of a problem. Look out for signs of recent re-undersealing (often carried out simply to hide well-established rust) and for areas which have been repaired by plating. Those vehicles which have spent their lives near the sea (and especially those used for towing boats into and out of the sea) tend to suffer most in this way, as the salt water promotes rapid corrosion. Vehicles which have been used in demanding off-road conditions may also bear the scars on their chassis - dents, scrapes and the like. If not attended to, scars like these can eventually become rust spots. While underneath a vehicle checking the chassis, it is also worth looking for evidence of leaks from the power steering box and hoses (where fitted) and from the transmission, engine and radiator.

On these early 110s, the bodies generally resist corrosion much better than chassis. However, their aluminium alloy panels will corrode where they meet steel panels, as a result of electrolytic action. Aluminium corrosion starts as bubbling under the paint and looks like a crumbly white powder when it breaks out. It is also true that the alloy panels are easily dented, and the fact that they will not normally corrode encourages owners to leave minor body damage unrepaired. As a result, many 110s

which have led a hard life will show every sign of it in their bodies.

While some of the body's steel cappings (such as the moulding on the doors and body sides at waist height) are galvanised and will therefore not corrode, others are not galvanised as they were on earlier Land Rovers and will therefore rust. Build quality was also patchy and one result of this is that some 110s suffer from rust in the windscreen pillars, where poorly seating windscreen seals have allowed water to collect and attack the steel.

Interior trim in all 110s of this age was very utilitarian, although it was rather better on County Station Wagons. In vehicles used for commercial purposes, it is very likely to have been damaged ; in Station Wagons, serious wear or damage on any except the front seats will suggest that the vehicle may have been used as a crew carrier for building or light industrial purposes, and has probably therefore been used roughly. Body leaks were very common on these early vehicles, and the interior trim may well bear witness to problems of this kind.

On the road, a 110 feels big and unwieldy to anyone not used to driving a 4 x 4. By comparison with a Series III Land Rover, however, it is a delight to drive ! Transmission snatch is normal (although serious clunks and knocking noises suggest excessive wear, probably in the propshaft UJs); gear whine from the transfer box is normal (and gets worse with age); and the four-speed gearbox fitted to V8 models is a rather unrefined component with a very vague gearchange at the best of times. Steering feels vague on all models, and the power-assisted type is well worth having because the manual steering box is heavy to use at parking speeds. The soft coil springs give excellent ride characteristics, but at the expense of considerable cornering roll.

The coil spring suspension comes into its own when the vehicle is driven over rough terrain, as it provides a vastly superior ride to the leaf springs of older Land Rovers. In soft or muddy terrain, it allows far greater axle articulation than leaf springs, and therefore gives each wheel a better chance of retaining traction over undulations. The permanent four-wheel drive fitted to most early 110s also reduces the risk of bogging, as it removes the need for the driver to judge when to engage drive to the front axle. However, it is worth remembering that a 110 is not best suited to some types of off-road work because its size makes it less manoeuvrable than a smaller vehicle. In addition, the long wheelbase increases the risk that it will belly-out over hummocks.

When checking the mechanical condition of a used 110, it is important to check that all the transmission components do function correctly. On those models with permanent four-wheel drive (which the vast majority did have), check that the differential lock works and that low range does engage and disengage correctly. Check that the handbrake will hold the vehicle on a gradient - but do

NOT check it by applying it on the move because it operates on the transmission and will certainly cause damage.

If the handling of a 110 feels particularly vague and woolly, the cause may well be worn steering or suspension bushes. One particular weakness is the Panhard rod bushes, but the whole coil-spring set-up depends heavily on rubber bushes for its correct functioning, and these bushes can all wear. It is worth considering replacing all the bushes as a matter of course on a high-mileage vehicle if there is no evidence that this has been done recently.

As far as engines are concerned, there are no major problems. The four-cylinder petrol and diesel types are robust and rugged, and seem to last forever. They are also easy to work on. On diesels, however, watch out for excessive smoking from the exhaust and for excessive oil consumption. The diesels are noisy and raucous and, like the four-cylinder petrol engines, can also suffer from noisy timing chains. This kind of noise is always worth investigating, because it is often caused by a chain which has lost its tension and is about to jump a sprocket and cause mayhem among valves and pistons.

The most refined and most powerful of the 110 engines is the petrol V8, which is another well-respected and durable piece of machinery, although its alarmingly high fuel consumption often deters buyers. Worn examples often suffer from top-end rattle, which generally comes from a worn camshaft or worn tappets, and can often be caused by an owner's failure to change the oil at the recommended intervals. The twin carburettors can also go out of tune, causing poor starting and rough running. Overheating suggests that there may be blockages in the waterways, often caused by aluminium corrosion in the engine if an owner has not used the correct coolant inhibitor.

THE CREW CAB AND
THE 2.5 DIESEL, 1984

This exploded view of the 2.5-litre diesel engine shows the toothed belt drive which had replaced the roller chain of the 2.25-litre engine.

The 110 continued in production unchanged until the end of 1983. In the meantime, it had been introduced into more and more overseas markets, and development work had continued at Solihull to strengthen the vehicle's market appeal. The Special Projects Department (later known as Special Vehicle Operations) had developed an ultra-long wheelbase twin-axle version and was working on a variety of other conversions, including a three-axle model, and the power train engineers had been working on a more powerful edition of the diesel engine.

Right from the beginning, sales catalogues for the 110 had advertised a "Crew Cab" model, available to special order only and clearly on a stretched wheelbase. The Crew Cab Land Rover was in fact not a line-built vehicle but was converted from a 110 by the Special Projects division of the company, and the first examples did not go to customers until the end of 1983.

The essence of the Crew Cab Land Rover was its longer wheelbase, stretched by 17 inches to give 127 inches between axle centres. The bodywork was designed to take a shortened High-Capacity Pick-Up load bed at the rear, together with a four-door, six-seater cab which was constructed from a mixture of Station Wagon and truck cab parts. The vehicle had the same 3,050 kg Gross Vehicle Weight as a standard 110, and could be ordered with any one of its three engines. As its 127 inch wheelbase clearly made the 110 name unsuitable, the Crew Cab Land

Rover was always fitted with the silver grille badge of the old V8 Stage I model, which read simply "Land-Rover". The hyphen in the name was an anachronism by this date of course.

The main purpose of the Crew Cab model was to provide both passenger-carrying space and load-carrying space in the same vehicle. It was thus ideal for applications where a team of workmen had to be transported along with their equipment. Crew Cab Land Rovers were bought by several public utility companies in the UK, such as Electricity Boards, but they were never very numerous.

It was in February 1984 that the next important change to the 110 range occurred, and this was announced at the Amsterdam Show which opened on the fourth of that month. The four-cylinder diesel engine was replaced by a new 2.5-litre (2,495cc) diesel derivative, which offered a 12% power increase (to 67bhp) and a 10% torque increase (to 113 lb/ft at 1,800rpm). Most important, from a driver's point of view, was that the new engine offered both improved acceleration and lower fuel consumption.

The new diesel had in fact been under development since shortly after the five-bearing versions of the older diesel had been introduced in 1980, and its increase in capacity had been achieved by fitting a longer stroke. This was hardly revolutionary: Land Rover had built an experimental 2.5-litre diesel engine in much the same way as long ago as 1962! However, what was new was that the

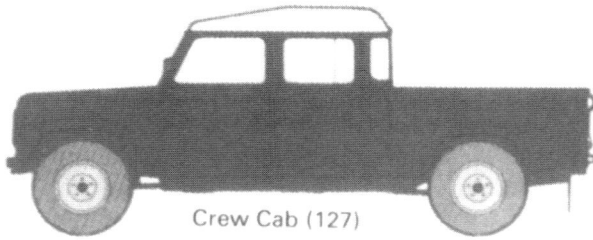

The Crew Cab model with its 127-inch wheelbase had a four-door cab and a shortened version of the High capacity Pick-Up bed. This illustration comes from Land Rover sales literature.

roller timing chains of the older engine had been replaced by a toothed rubber belt of the type which had become popular in recent years. This belt also drove the new DPS injector pump which was designed to give more accurate fuel metering, to retain its tune for longer than earlier types, and to minimise fuel consumption. It also benefitted from a self-priming system instead of the hand primer of earlier types. Cold-weather starting in the new engine was also improved by means of more efficient sheathed-element glow plugs, and the alloy castings used for its water pump housing and front cover also saved weight.

"There will not be a 2 1/2 litre petrol engine," claimed Land Rover's own Product Bulletin of December 1983, which advised dealers of the forthcoming new engine. As time would tell, however, that emphatic statement was misleading.

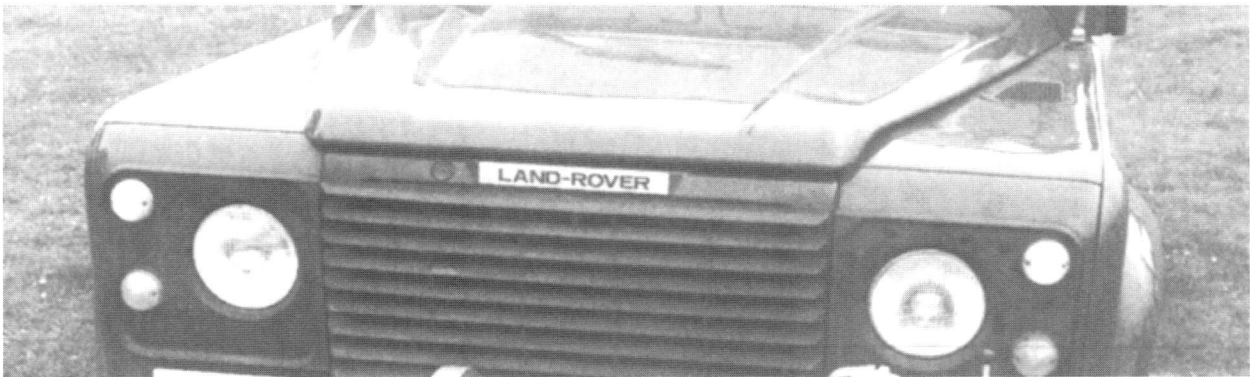

Crew Cab models had the silver grille badge first seen on Stage I V8 models in 1979.

B U Y I N G   A   1 9 8 4   M O D E L

The regular 1984 model 110 Land Rovers had the same strengths and weaknesses as the first 1983 models, with the single exception that later examples had the enlarged diesel engine. This is discussed below. In Great Britain, the 1984 models were distinguished by A-prefix registrations.

By this stage it was apparent that Station Wagon variants of the Land Rover 110 were selling very much better on the home market than the older leaf-sprung 109 variants had. It was also clear that the utility variants were selling less well, as there were cheaper vehicles to be had from other manufacturers. A large proportion of 1984 model 110s sold in Great Britain were Station Wagons, and of these probably the majority were County models. This sales pattern is obviously reflected today in the availability of used vehicles from this period.

However, the arrival of the Crew Cab model with its 127-inch wheelbase allowed Land Rover to offer a utility vehicle in a market sector which other manufacturers had not yet tackled. It was a small sector, and the Crew Cab models were not strong sellers, but they did have the advantage of persuading some fleet users who needed their extra carrying capacity to buy fleets of mechanically similar 110s as well.

All Crew Cab vehicles were sold for commercial purposes and they had little appeal to the private buyer. However, used examples could suit small businessmen who need to carry several employees and some equipment in the same vehicle. Their major drawbacks are heavy fuel consumption, size, and an even greater tendency to wallow on corners than their 110 counterparts. Buyers who are certain that they want one are not likely to find examples for sale very easily. Mechanically and bodily, the Crew Cab has the same strengths and weaknesses as the contemporary 110.

The 2.5-litre diesel engine offers a worthwhile improvement in road performance over the superseded 2.25-litre type, although it still leaves a 110 (and even more so a Crew Cab) feeling somewhat underpowered. It is almost as noisy as the older diesel and many owners find it hard to detect any decrease in noise levels as a result of the toothed belt drive instead of the roller chain. Many of the 2.5-litre diesel engines supplied to the British Army in the mid-1980s developed faults which were rectified under warranty, but the enlarged diesel does not have a particular history of problems in Land Rovers.

Nevertheless, the rubber timing belt does stretch after prolonged periods of use and Land Rover recommend that it is replaced periodically to prevent timing problems. Some owners consider the recommended replacement intervals leave enough room for problems to develop undetected, and have preferred to replace the belt by Zeus timing gears, an aftermarket conversion.

THE 90 AND THE 1985-MODEL 110

Land Rover Ltd have only rarely stuck to the traditional autumn introduction date for new models. The 110 arrived in March 1983; the 2.5-litre diesel engine appeared in February 1984 and the next developments occurred in June 1984. The changes they brought created what were widely seen as the 1985-model Land Rovers. The 110s were improved in a number of areas and the anticipated new short-wheelbase model was announced with the designation of Land Rover 90.

The 90 name was in fact rather misleading, as the new short-wheelbase Land Rover had a wheelbase of 92.9 inches or, more precisely, 2,360mm in the metric standards to which Land Rover now worked. Early prototypes did in fact have a 90 inch wheelbase, but this had been discarded as unsatisfactory before production began.

Like the 110, the 90 was characterised by coil sprung suspension and disc front brakes, and it shared the larger model's styling cues. Also like the 110, it was expected in due course to replace its Series III equivalent, although for the moment the 88-inch models remained in production. On the home market, however, the 88-inch models more or less disappeared from the showrooms as soon as the 90s became available.

Compared with the 88-inch Series III, the 90 was 4.4 inches longer overall and, like the 88, it had a shorter rear overhang than its long-wheelbase cousin. This had the advantage of improving the departure angle for off-road driving, and in addition it placed the wheels almost ideally at each corner of the vehicle to give the best possible ride. Meanwhile, the extra 4.9 inches in the wheelbase as compared to the Series III 88 had allowed Land Rover to provide more cargo space and improved front seat travel. Payload too was increased by between 14% and 33%, depending on the body type. As with the Series III 88, four different bodies were built on the assembly lines, these being the soft-top, hardtop, pick-up and seven-seater Station Wagon. Station Wagons could be ordered with County trim, and the 90 was also available to special order as a chassis/cab for special bodywork.

Just as the Series III 88 had been available only with the four-cylinder petrol and diesel engines, so there was as yet no V8 petrol engine for the 90. The engine choice rested between the 74bhp 2,286cc petrol type and the new 67bhp 2,495cc diesel. As on the 110 models, these came with the five speed LT77 gearbox and LT230R transfer box, although the lighter weights of the short-wheelbase models had enabled the Land Rover engineers to specify a taller and more fuel-efficient "High" ratio in their transfer boxes. However, permanent four-wheel drive was mandatory.

The switch to permanent four-wheel drive had also been made for the four-cylinder 110s, because the selectable four-wheel drive option had found few takers. However, this was the only mechanical change which affected the facelifted 110s announced in June 1984; all

the other changes were to equipment levels. And, of course, the changes made to the 110s were also part of the launch specification of the 90s, where appropriate.

It was very easy to distinguish one of the facelifted 110s from an early model at a distance because the newer models had single-piece glass in their front doors. Early 110s had detachable door tops with sliding windows, exactly like their Series III counterparts, but these were rather crude devices - particularly for a vehicle of the County Station Wagon's cost and pretensions - and so a decision had been made to equip the facelifted models with winding windows. For the moment, however, the doors were still made in two pieces, with the join disguised by a galvanised garnish rail, just as it always had been.

With the new window arrangements came smart moulded interior trims incorporating locking buttons on their sills. Station Wagon bodies also had modified sliding windows in the rear, which tidied up their appearance. Soft tops and three-quarter tilts were now supplied in Matt Stone instead of Khaki, presumably because the new colour had a rather less military appearance; needless to say, Khaki soft tops were still available for military buyers, though!

Standard equipment levels had also been increased, and all models now came with a number of items which would have been considered the height of luxury on Land Rovers only a few years earlier. The reason, of course, was that more car-like utility vehicles - particularly from Toyota - had changed customer expectations in this sector of the market. From June 1984, all 110s - and 90s, of course - came with a cigar lighter, a steering column lock, a bonnet lock, a fuel filler lock, and a dipping interior mirror. Station Wagons now came as standard with a heated rear window and rear wash-wipe, both of which were very welcome additions.

As before, County Station Wagons headed the range, and they were available in both 90 and 110 forms. The list of standard equipment was lengthened slightly by the addition of twin radio speakers and an aerial (the customer had to choose a radio or radio/cassette and pay extra for it), and by the addition of carpeting in the passenger area. There were new side decals, too, this time colour co-ordinated with the main body colour and themselves in two tones with the "County" name emblazoned across the front doors.

During this period, Land Rover Ltd were in the process of centralising their manufacturing operations at the Solihull site, and of closing down most of the smaller factories which supplied components used in the assembly of Land Rovers. At the same time, they were redeveloping existing buildings at Solihull to give maximum efficiency and production flexibility. As part of this redevelopment, the South Works had now been equipped with three assembly lines for coil-sprung Land Rovers instead of the one on which 110 assembly had started at the beginning

When it appeared, the 90 actually had a wheelbase of 92.9 inches. Styling presented no surprises after the 110. This is one of the first, a County Station Wagon.

The facelifted 110s were much tidier-looking than the original versions. This County Station Wagon shows off the new side decals associated with the 1985 models and the neater rear window with its rounded corners. Note also that the upper body sides were no longer in Ivory, which was now reserved for the roof.

of 1983. One of these was dedicated to 90s, one to 110s, and the third was designed to deal with either model, so that the ratio of short-wheelbase to long-wheelbase models could easily be adjusted to suit demand.

Of course, not all coil-sprung Land Rovers were now being assembled at Solihull. CKD assembly had started overseas from kits of parts shipped out from the Land Rover factory, and among the more notable of the overseas assembly plants was the new one belonging to Jaguar-Rover Australia at Moorebank in Sydney. From November

1984, that plant began to assemble 110 models - and it would not be long before it would also begin to produce some rather special variants of its own.

The first Australian-built 110s were all County Station Wagons, and these differed from models made at Solihull in a number of respects. Only two engines were available, one a high-compression (Range Rover tune) version of the petrol V8, and the other a 3.9-litre Isuzu diesel which is described in detail elsewhere. In addition, the Australian-built 110s had a rear anti-roll bar as standard.

THE V8 90 AND 2.5 PETROL ENGINE FOR 1986

After the 90 had arrived in June 1984, Land Rover waited a further 11 months before announcing the model which so many customers had been waiting for. The V8 90 was introduced in May 1985, once again not coinciding with the beginning of the traditional model-year. However, Land Rover did manage to keep things tidy for the second major announcement of 1985, and the 2.5-litre petrol engine which many people had expected (despite the company's outright denial that there would ever be such an engine) appeared in August 1985, just a few weeks before the autumn Motor Shows which normally begin the European model-year.

The V8-powered 90 was the first factory-produced short-wheelbase Land Rover with the V8 engine, although there had been many private conversions before this and the Land Rover engineers had in fact built their first experimental short-wheelbase V8 in the mid-1960s. In the V8 90, the engine had the same 114bhp, 185 lb/ft tune as in the 110, but it was now equipped with electronic ignition to help maintain tune for longer periods and to give easier starting. Electronic ignition was also added to the V8 engines in 110s built after May 1985.

The light weight of the 90 made it very much quicker on the road than its long-wheelbase cousin. In fact, a 90 V8 was slightly quicker to 60mph from a standing start than the contemporary carburetted Range Rover; the latest fuel-injected Range Rover Vogue models nevertheless ensured that the more expensive vehicle could still lay claim to the best performance in Solihull's range.

One of the less desirable features of the V8-powered 110s had been their Range Rover four-speed gearboxes, because the slow and deliberate changes these demanded took the edge off acceleration. Solihull took the opportunity to put this right with the introduction of the 90 V8. Even though the LT77 five-speed box had been used successfully with the V8 in Range Rovers since autumn 1983, the Land Rover engineers doubted whether it would be strong enough to withstand both V8 torque and the heavy payloads which utility versions of the Land Rover were often called upon to carry. So it was that they had chosen a different five-speed gearbox for the V8 Land Rovers. This was the LT85, a heavy-duty type which had

been designed in Britain but was built by Santana, Land Rover's Spanish affiliate. The only Solihull-built Land Rovers to which it was fitted were the V8-engined 90s and 110s; four-cylinder models retained their LT77 five-speed gearboxes.

Over the summer of 1985, the last of the Series III 88 inch and 109 inch models were built, and at the same time the old 2.3 litre four-cylinder petrol engine ceased production. In August 1985, this was replaced by a more powerful 2.5-litre version which embodied most of the same improvements as the 2.5-litre diesel, with the same lengthened stroke to take the capacity up to 2,495cc, a similar toothed belt auxiliary drive in place of the original roller chain, and lightweight castings. As they had on the diesel engine, these changes gave the four-cylinder petrol 90s and 110s better acceleration and better high-speed cruising ability. Power was now up by around 12% to 83bhp, while torque had increased by nearly 11% to 133 lb/ft, at 2,000rpm instead of the 1,800rpm of the older engine.

The net result of all these changes was that, after autumn 1985, all Land Rovers built at Solihull could be had with 2.5-litre diesel, 2.5-litre petrol or 3.5-litre petrol engines, and that all of them came as standard with five-speed main gearboxes and permanent four-wheel drive. There was some time lag before all CKD models built abroad also had all the new mechanical components, and the five-speed LT85 gearboxes did not become available on Australian-built 110s, for example, until the end of the year. However, the last vehicles to be brought into line were the Australian-built 110s with Isuzu diesel engines, which were given LT85 five-speed gearboxes in February 1986.

Meanwhile, Land Rover's Special Projects Department had been reorganised, and from July 1985 it became known as Special Vehicle Operations, or SVO for short. Building on the success of the long-wheelbase Crew Cab Land Rover, it now began to offer the 127-inch chassis with a limited range of alternative bodies and as a chassis/cab for Land Rover's approved converters. The alternative bodies from SVO were a low-line box-body to go with the crew cab, and a high-roof box body to go with a standard two- or three-seater truck cab. These vehicles first became available in the early part of 1986, and with their introduction the name Crew Cab Land Rover was obviously no longer appropriate. As a result, the extra-long wheelbase Land Rover became known as the Land Rover 127, although it retained its plain "Land-Rover" grille badge (still with the hyphen) for the time being.

The first sales leaflet for the V8-engined 90 in the summer of 1985 featured this "ghost" picture of a County Station Wagon.

BUYING A 1985 OR 1986 MODEL 90 OR 110

The first facelifted 110s and the first 90s appeared early enough to attract A-prefix registrations in Great Britain, but the majority of the 1985 models had the B-prefix numbers which were issued from 1st August 1984. Similarly, the first V8 90s and the first 2.5-litre petrol engined 90s and 110s had B-prefix registrations, but the majority of the 1986 models were registered with the C-prefix numbers issued from 1st August 1985.

The facelifted 110s introduced in June 1984 were very much more civilised and refined vehicles than their predecessors, not least because of their new doors with winding windows. Higher specification levels on these models also played their part in making all models - and particularly the popular County Station Wagons - easier for the car-orientated owner to live with. The new doors and higher equipment levels were of course standard on all 90s. However, the new door trims on all types did incorporate one annoying design fault: their sill locking buttons tend to foul the driver's elbow, with the result that drivers often find they have locked themselves in accidentally.

The 90 was an immediate success, offering a civilised and credible alternative to a car which did not have the handicap of the 110's size. Many Station Wagon versions were bought for their fashionable appeal rather than with any intention that they should be used for rough-terrain work or for towing. The arrival of the V8 version with its much better performance enhanced this chic appeal even further, and even the vehicle's heavy fuel consumption did not prevent it from selling strongly. However, in retrospect, these early V8 90s are the least desirable of their type, because they have the lower-powered (114bhp) engine. Not until the 1987 models were announced in October 1986 did the higher-powered (134bhp) engine become available.

All the V8-engined 90s and all 110s built after the 90 was introduced had the new Santana-built LT85 five-speed gearbox. Opinions vary about this component: some owners argue that it offers a better change quality than the Rover-built LT77 five-speed, while others insist that it feels very agricultural. One thing is certain, however, and that is that the LT85 is very much perferable to the old LT95 four-speed fitted to the earliest 110s. Its change quality is very much better, and the fifth gear does afford some improvement in fuel economy at cruising speeds.

The 2.5-litre petrol engine introduced in August 1985 did offer a performance advantage over the old 2.25-litre type, but it still did not make even a 90 into a good road car. As it offered neither the economy of the diesel nor the road performance of the V8 petrol engine, the 2.5-litre petrol engine was the least popular of the 1986 options and is correspondingly scarce on the used market today.

For buyers who intend to use their vehicles for demanding recreational off-roading, a 90 is undoubtedly the best buy among Land Rovers of this vintage. Either petrol V8 or diesel engines will do (and it will usually be road use which determines the choice), because both offer excellent off-road performance. The 90's short wheelbase makes it much less likely to belly-out on humps than a 110, and also makes it that important bit more manoeuvrable. Even today - ten years after the 90 was first announced - it is still widely respected as the best dual-purpose off-roader on the utility market, new or used.

As before, the ultra-long wheelbase model was available in limited numbers and was still for commercial use only. There will be little choice for buyers, and the 127s have very little appeal except to commercial users.

1 9 8 7   M O D E L S :   T H E   D I E S E L   T U R B O
A N D   U P R A T E D   V 8

The two key exterior changes for 1987 were the single-piece doors and the new door handles. Also visible here are the 1987-season County side decals, the third style used in just over four years of production.

Changes came thick and fast in the mid-1980s and, with the V8 90 and 2.5-litre petrol engine only just beginning to make an impact on sales, Land Rover announced further changes to its 90 and 110 ranges in October 1986. Principal among these were increases in power for the V8 petrol engines and for the diesel - the latter by means of an optional turbocharger - but there were also important interior and exterior changes to the vehicles. These changes came only just in time. Sales had been falling badly since 1980 and Land Rover's total vehicle sales for 1986 were barely more than half those for 1980, while annual sales of the 90s and 110s had fallen by some 30% since their introduction in 1983. But 1986 proved to be the turning point: during 1987, 90 and 110 sales increased for the first time.

The main problem which the 90s and 110s had been facing was strong competition from Japanese vehicles. In the utility market, Toyota Hi-Lux , Nissan and Isuzu pick-ups all offered greater refinement, albeit without the ultimate rough-terrain ability or the long-term durability of the Land Rover products. In the passenger-carrying market, the Toyota Land Cruiser, Nissan Patrol and Isuzu Trooper promised much of the appeal of the more expensive Range Rover with greater refinement and lower running costs than a 90 or 110 Station Wagon. These,

then, were the problems which Land Rover set out to address with the October 1986 changes.

Two changes to the exterior were immediately visible. The first was that the traditional recessed door latches had been replaced by car-like handles with push-button releases which were much easier to use and less likely to injure the user's hands; and the second was that there were now one-piece doors in place of the traditional two-piece type with detachable tops and a garnish rail to hide the join. Nevertheless, the garnish rail remained in place on the rear flanks below the side windows. The one-piece doors brought with them redesigned interior panels.

County Station Wagons, on both the short and long wheelbases, were also dressed up with new and more attractive side decals, with the County name once again on the front doors. This change was principally intended to give the vehicles a smarter and more fashionable appearance. For similar reasons, 90 County models were fitted with enamelled Rostyle pressed-steel wheels (with the five-spoke pattern used on Range Rovers since 1970), although the 110s retained their plain disc wheels.

The redesigned door trim panels had recessed grab handles and gave the interior a much less utilitarian look than before, and they incorporated flush-fitting locking buttons which were both harder for thieves to operate from

Land Rover held over the 1987 models until December 1987, and so many of them were registered in the UK with the E-prefix registrations generally associated with 1988 models. This press picture of a 110 County Station Wagon was released at the time of the Motor Show in autumn 1987.

outside and no longer got in the way of the driver's elbow. There was further refinement in the dashboard, which now incorporated provision for a radio in its centre; and there was an internal bonnet release operated by a cable, which replaced the catch on the utility models and the lock on the County Station Wagons. A particularly welcome addition - although not one which was readily visible - was that the rearmost door on Station Wagons now incorporated a stay which would hold it open at right angles to the body.

All these changes were designed to meet the mid-1980s customer expectations that a 4x4 should be more like a car to use than like a van or light utility vehicle, and so it was not surprising that they should be accompanied by mechanical changes which made the 90s and 110s more acceptable for everyday use by those who were used to the characteristics of a car. While the four-cylinder, 2.5-litre, petrol and diesel engines remained unchanged, the V8 3.5-litre petrol engine was uprated and a turbocharged version of the four-cylinder diesel engine was introduced to offer acceptable road performance in tandem with diesel fuel economy. By this stage, diesel engines had come to dominate the 4x4 passenger market in continental Europe (although things were different in diesel-shy Britain), and it was important for the 90 and 110 to offer competitive performance - which the old, naturally-aspirated diesel engine clearly could not. For exactly the same reasons, the Range Rover had been equipped for the first time with a turbocharged diesel engine just a few months earlier, in April 1986.

The main effect of the changes to the V8 petrol engine was to improve the high-speed cruising ability of the 90s and 110s to which it was fitted. A new camshaft was largely responsible for increasing maximum power from 114bhp at 4,000rpm to 134bhp at 5,000rpm and twin SU carburettors replaced the original Zenith-Strombergs to give slightly better fuel economy. Torque too, was increased, although only marginally to 187 lb/ft from the earlier 185 lb/ft at the same crankshaft speed of 2,500rpm.

The turbodiesel engine was always marketed under the name of Diesel Turbo, which was Land Rover's way of distinguishing it from the completely different turbodiesel engine in the Range Rover Turbo D. Whereas the Range Rover engine was made by VM in Italy, the Land Rover Diesel Turbo engine was built at Solihull, and was a derivative of the existing four-cylinder 2.5-litre engine. Development had taken two years, and had concentrated on increasing bottom-end torque to give good acceleration and on minimising turbo lag - the time delay between pressure on the accelerator pedal and the turbocharger cutting in to increase acceleration. Maximum torque had been increased by 28%, while power had also gone up by 25% to give a noticeable improvement in high-speed cruising ability. The downside of all this, however, was that the Diesel Turbo engine used more fuel than its naturally aspirated parent.

There were some very close links between the new Diesel Turbo engine and the old 2.5-litre diesel; the belt-driven overhead-valve configuration was the same, and the bore and stroke dimensions were identical. However,

Another autumn 1987 Motor Show press release picture, this time showing a 90 County with the enamelled Rostyle wheels introduced in October 1986. Note also the "V8" identifier at the trailing edge of the side graphics. The protruding grille panel allows room for the additional radiator and cooling fans necessary when the optional air conditioning system was fitted.

there were also many differences. The turbocharged engine had a completely new cylinder block, which incorporated an oil feed and drain for the turbocharger. Its crankshaft, too, had been revised and now had cross-drillings to improve the lubrication of the bearing journals. There were new pistons and new piston rings, and nimonic exhaust valves had been specified to cope with the higher combustion temperatures associated with turbocharged engines. For the same reason, the water cooling system had been uprated and fitted with a viscous-coupled fan, and an oil cooler was now standard equipment.

The turbocharger itself was a Garrett AiResearch T2 model, with integral wastegate limiting the boost to 10psi. The fuel pump was a self-priming DPS type with a boost control capsule and a cold start timing retard device, and weight had been saved by a new alloy-bodied vacuum pump for the brake servo and by a lightweight starter motor which also gave higher cranking speeds. Lastly, there was a new air cleaner, which breathed through an intake on the left-hand front wing. This intake is an easy way of distinguishing a Diesel Turbo from other models of the 90 and 110 at a distance.

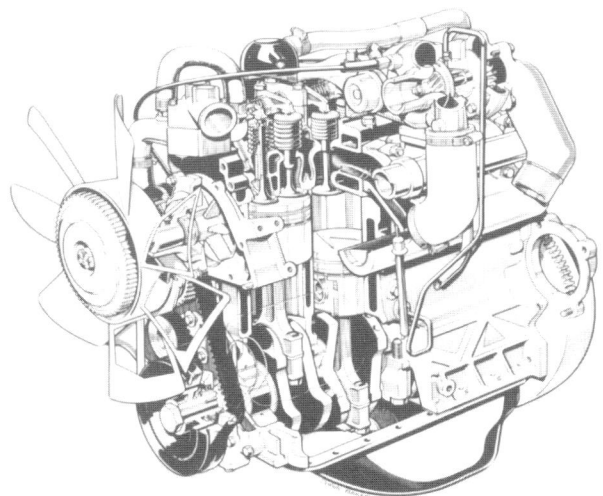

This cutaway drawing shows the essential features of the Diesel Turbo engine: the turbocharger is mounted high up on the nearside.

BUYING A 1987 MODEL

Home market 1987 models, which included the Diesel Turbo and 134bhp V8 petrol engines, mostly had the D-prefix registration numbers issued between August 1986 and the end of July 1987. However, as the 1988 models were not made available until the very end of the 1987 calendar year, a number of 1987 models also had E-prefix registrations.

The one-piece doors on 1987 models not only improve the appearance of 90s and 110s by dispensing with the galvanised garnish rail at waist height, but also provide better wind- and weather-proofing than the older types. In use, their new exterior handles are a vast improvement over the earlier recessed type. They are less prone to rattles and squeaks, too, and this adds to refinement levels. On the Station Wagons, the improved rear door stay makes much more of a difference in use than might be expected.

The 134bhp V8 engines do make the 1987 and later models to which they are fitted better high-speed cruisers than examples with the earlier 114bhp V8; but they are also even thirstier. While they remain the best engines to choose for road use, the new Diesel Turbo is also worth considering because it is so much more powerful than the earlier naturally aspirated types. However, fuel economy is no better - usually rather worse because owners tend to use the new-found performance to the full - and noise levels and general refinement are still a source of irritation to many buyers.

Despite rumours to the contrary, the Diesel Turbo engines are no less reliable than the naturally aspirated types, provided that they are properly maintained. Rumours of weakness originated when some owners, used to maltreating an older diesel and changing the oil only when they remembered, discovered that this kind of treatment would cause the turbocharger to run its bearings. The fact is that the turbocharger is lubricated from the engine oil supply and runs at very high revolutions: it will not tolerate dirty oil or an insufficient supply of lubricant. Buyers should therefore remember that the engine oil must be changed at the recommended intervals and that it is best not to attempt to economise by buying cheap oil. Also advisable is not to switch off an engine which has been running at high speed without first allowing it to idle briefly. There is a risk that the turbocharger may still be spinning and that the sudden oil starvation when the engine is switched off could lead to bearing damage.

THE FACELIFT FOR 1988

The County Station Wagons were once again the style leaders of the range for 1988. This is a 110 County, with the new sunhatch, body-coloured front panel, black bumper and latest side graphics. The registration number shows that this early example was registered by the factory very shortly after the 1987 models illustrated in the previous section.

The roller-coaster of changes to the 90s, 110s and 127s continued for the 1988 models, which were announced at the Royal Smithfield Show on 6th December 1987. The new season's models brought mainly cosmetic differences, and the main focus was on the County Station Wagons with which Land Rover hoped to strengthen their presence in the market sector to which they planned to introduce the still secret Discovery in autumn 1989. However, the naturally-aspirated diesel engine was relegated to an export-only option, leaving only the 2.5-litre petrol, 3.5-litre petrol and 2.5-litre Diesel Turbo engine options available on the home market. In practice, the older diesel engine was available only in underdeveloped countries where the additional complication of the turbocharger installation was likely to cause maintenance problems.

All the 1988 models had black bumpers in place of the galvanised-finish variety which had been standard wear since the very first Land Rovers had been built in 1948. On the workhorse models, these were complemented by a black grille, black headlamp surround panels and black wheelarch eyebrows; on County Station Wagons, however, the grille, headlamp surrounds and wheelarch eyebrows were all finished in body colour.

The County Models also picked up new side stripes

yet again - the fourth variety since 1983 - and a new Britax tilt-and-remove sunhatch. This had a dot-screened glass panel which could be tilted into one of three positions or taken out completely. The sunhatch could also be ordered as an extra-cost option on other Land Rover models, and Land Rover Parts and Equipment offered an accessory pack to go with it. This consisted of a wind deflector, a storage bag for the glass panel and a clip-on interior sunblind which also acted as an insulation and trim panel.

Other, more minor exterior changes distinguished the 1988-model Land Rovers from their predecessors. Arrow Red and Shire Blue were added to the paint options list, although, as they replaced three colours from the 1987 range, the total number of paint options actually went down from eight to seven. At the rear, V8 petrol and Diesel Turbo Land Rovers now wore silver identifying decals - "V8" for the 3.5-litre petrol vehicles and "Turbo" for the Diesel Turbos. Four-cylinder petrol and naturally-aspirated diesel models however had no identifying decals at the rear.

Interior changes gave the 1988 Land Rovers an even more car-like ambience. Grey seat facings were complemented by dark grey door trim panels, and carpets and headlinings, when fitted, were in a lighter grey. The

For 1988, the utility models retained the black grille and headlamp surround panels, but also had black wheelarch eyebrows. From the beginning, the extra width of the HCPU body had meant that eyebrows were not fitted to the rear wheelarches.

Utility models for 1988 still had vinyl seats like these, with a cubby box in the centre as standard.

1988-model County trim was in grey cloth. Home market Station Wagons normally came with a third seat at the front, as shown here.

All the 1988 models had this revised dashboard, with provision for a radio in the centre. The steering wheel boss, too, was slightly different.

Land Rover used this appropriately-registered 90 Soft Top in their 40th Anniversary celebrations during April 1988, but the body-coloured wheelarch eyebrows show that it was actually an obsolete 1987 model !

steering wheel boss was mildly altered and now had horizontal styling lines above and below the Land Rover name instead of a recessed centre panel. Its rim was trimmed with leather on County Station Wagons, which also came with a stereo radio-cassette player as standard equipment. Last, but by no means least, there were improved window seals for all models, which closed over the windows when these had been wound down into the doors and thus sealed the door cavities more effectively.

During April 1988, the Land Rover marque celebrated 40 years of continuous production and a new logo was adopted to coincide with the anniversary. The old green and yellow colours were replaced by a less garish green and cream combination and the oval badge changed to suit, gaining a cream border at the same time. Nevertheless, this change did not yet affect the 90, 110 or 127 vehicles.

B U Y I N G   A   1 9 8 8   M O D E L

The earliest 1988 facelifted models had E-prefix registration numbers, but later examples had the F-prefix numbers issued after 1st August 1988. On the home market, a majority were equipped with either the V8 petrol engine or Diesel Turbo; the naturally-aspirated diesel was no longer offered through the showrooms and the ageing four-cylinder petrol engine was not a popular choice.

As the 1988 models differed from their predecessors only in cosmetic details, there is little to distinguish them from earlier examples. The sunhatch is certainly worth having and the neater interiors do provide an even more car-like ambience than before. However neither feature is worth paying very much extra for in a used example.

THE 1989 - 1990 SEASON MODELS

Unchanged graphics on the 90 County Station Wagon made it hard to distinguish a 1989 model from a 1988 at first glance.

The 1989 model-year revisions to the 90s, 110s and 127s were announced right at the end of 1988, and the venue chosen was once again the Royal Smithfield Show, which this year opened on 4th December. Although the changes were few in number, they represented an important step in the process of refinement which would turn the coil-sprung Land Rovers into Land Rover Defenders some 20 months later.

There were further revisions to the paint options for all models and side stripes now became standard across the 90 and 110 range: County Station Wagons retained those introduced in December 1988, but the utility models had new stripes which incorporated a large "90" or "110" identifier. The 127-inch models, however, were delivered without side stripes. Rostyle wheels, already standard on 90 County Station Wagons, now became an extra-cost option for all 90s. And optional on the cheaper vehicles but standard on both 90 and 110 County models was a new radio-cassette which incorporated a Long Wave receiver and pre-set facility for 16 stations.

Probably the most noticeable change for the 1989-season Land Rovers affected only the Hard Top and Station Wagon versions. The rivet-heads in the upper body sides of these vehicles had always been a reminder of the vehicle's essentially utilitarian nature and had always seemed a little out of place on vehicles with the pretensions to refinement of a County Station Wagon. So for 1989 the upper body sides were redesigned to leave no rivet-heads exposed, a refinement which left the vehicles looking much less utilitarian.

In addition, 90 Hard Tops could now be ordered with an optional Roof Appointment Pack - a rather grandiose name for what was no more than the complete roof and trimmings from the 90 County Station Wagon. It included Alpine lights, the tilt-and-remove sunhatch and the full Station Wagon style of headlining.

There were no further major changes for the 1990 model-year, which began at Motor Show time in October 1989. However, 1990 models were distinguished by new grille badging: the new corporate oval Land Rover logo was offset to one side of the grille itself and the original model badges were replaced by "90" or "110" badges as appropriate.

The 1989-season utitlity models were easy to distinguish, however, thanks to their use of side graphics for the first time.

Optional enamelled Rostyle wheels brightened up the appearance of even the most utilitarian of Ninetys. This is a pick-up model, with truck cab and tilt.

The 1990-model vehicles had revised grille badges, as seen on this 110.

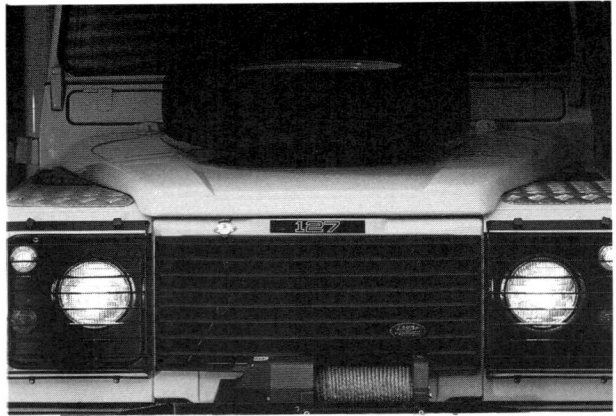

Rather a rarity is this 1990-model 127, showing the grille badge used only for that year.

BUYING A 1989 OR 1990 MODEL

Most 1989 model 90s, 110s and 127s had F-prefix registrations in Great Britain and the final 1990 models had G-prefix registrations. There was a certain amount of overlap between August and October 1989, when the outgoing 1989 models took on G-prefix numbers and so a G-prefix is not an infallible indicator that a vehicle has the 1990 specification.

In fact, there is nothing about the 1990 specification which is really worth looking for as it consists only of badging changes to the 1989 models. However, it is quite possible to make a 1989 model look like a 1990 model simply by changing the badges on the grille and front panel, so buyers should beware of 1989 models being passed off as later types and priced accordingly.

Both 1989 and 1990 models share the rivet-free upper body sides, which give a neater appearance but bring no practical benefits. From the point of view of everyday use, there is so little to distinguish one of these final 90s or 110s from the 1988 models that it is not worth paying extra for the later versions. Even buyers concerned about the advertisement of a vehicle's age contained in the registration number prefix could probably save money by buying an earlier vehicle and equipping it with a cheap undated number from one of the specialist registration number agencies.

As before, the vast majority of vehicles of this vintage have either the V8 petrol engine or the Diesel Turbo. Station Wagons were mostly built to County specification for the home market and they are far more plentiful than utility variants. The ultra-long-wheelbase 127 is rare and is once again found almost exclusively with specialist commercial bodywork.

ENTER THE DEFENDER FOR 1991

The 90, 110 and 127 Land Rovers were replaced in September 1990 by the Land Rover Defender 90, 110 and 130. In fact, the new vehicles were broadly the same as those they replaced, the only major differences lying in their drivetrains. The new 200 Tdi direct-injection diesel engine first seen in the Discovery a year earlier replaced both of the older diesel engines; the LT77 five-speed gearbox replaced the Santana-built LT85 behind V8 engines; and the four-cylinder petrol engine was relegated to a special-order option for the home market. Power-assisted steering was made standard on all models. There were also some cosmetic revisions which followed the trend set during the 1990s of making these essentially utilitarian vehicles more refined and car-like to drive.

Thereafter, the Defender range followed its own evolutionary path during the 1990s.

MILITARY 90s, 110s AND 127s

The Land Rover was already a favourite with military and paramilitary forces all over the world when the coil-sprung models were introduced in the early 1980s and examples of the new types soon found their way into service. Land Rover had prepared 110 military demonstrators shortly after the vehicle went on sale in 1983, but the first deliveries of military variants were not made until 1985. These were 110s, which went to the British Army. The earliest order from overseas was received in 1984 and by the end of the decade the 90 and 110 were in service with military and paramilitary forces all over the world. In addition, several military authorities took examples of the 127.

Foreign military authorities took coil sprung Land Rovers with all types of bodywork, some specially created to meet their individual needs. All types of engines have also been supplied for military use. However, the British Armed Forces standardised as far as possible on the naturally-aspirated 2.5-litre diesel, taking only a handful of 2.5-litre and 3.5-litre V8 petrol models for special duties. British 90s and 110s came as Soft Tops and Hard Tops and as Window Hard Tops and Station Wagons in smaller numbers for special duties. They were equipped either as GS (General Service) vehicles or as FFR (Fitted For Radio) types with 24-volt electrical systems to operate communications equipment in the field. In due course, no doubt namy of them will be sold on to the civilian market in the same way as their predecessors were.

It was the British Army which was the unfortunate recipient of a batch of Land Rovers with faulty diesel engines in 1985-1986. The problem was identified as incorrect manufacture of pistons, compounded by inadequate servicing in the field by soldiers for whom these were the first diesel-engined vehicles they had been required to maintain. All the engines affected were dealt with under warranty and the problem has never recurred, but it did cast a shadow over the reliability of the 2.5-litre diesel engine during the second half of the 1980s.

As the pictures which follow demonstrate, some of the military vehicles based on the coil-sprung Land Rovers bear very little resemblance to the standard civilian article.

This early demonstrator or prototype 110 military soft top was pictured at Solihull in May 1983. The side hatch ahead of the wheelarch was designed to accommodate a jerrycan.

This left-hand drive Diesel Turbo 90 Station Wagon was supplied to an overseas military customer in the later 1980s.

Another Diesel Turbo model for an overseas customer, this is a 110 Window Hard Top.

Based on a 90, this Solihull demonstrator is a gun platform. A large cannon would be mounted in the load area with its barrel pointing forwards and emerging between the two halves of the windscreen. On the bonnet is a blast shield which could be raised to protect driver and gunner when the gun was fired.

This Long-range Desert Patrol Vehicle was based on a 110. Similar vehicles were taken into service by the British SAS Regiment to replace their long-serving Series IIA "Pink Panther" 109-inch Land Rovers.

The 127 chassis also found favour with some military users. This Diesel Turbo ambulance was built for an overseas customer.

This Shorland Series 5 armoured car bears little resemblance to the 110 on which it was based. The original Shorland - built by Short Brothers and Harland in Belfast - was introduced on the Series IIA Land Rover chassis in the mid-1960s.

An alternative to the Shorland was the Hornet, built by Glover's of Hamble. Front end details immediately indentify the basic vehicle as a Land Rover 110.

Hardly recognisable as a Land Rover at all, the Hotspur-built Hussar used Hotspur's own 6x6 version of the 110 chassis with 150 inches between the first pair of axles. It was designed as an armoured personnel carrier and could carry 12 men; there was also provision for vehicle-mounted weapons in the revolving turret. The V8 petrol engine was standard.

AUSTRALIAN PECULIARITIES

The Australian market has its own peculiarities, and demanded some very special variants of the 90 and 110 Land Rovers which were not made available in other markets. Among those were variants with special engines, variants with special bodies and variants with special chassis.

By the end of the 1970s, the Australian light utility market was dominated by vehicles with diesel engines and Toyota were having a runaway success with their diesel four-wheel-drives. Against the power and torque of Toyota's big six-cylinder diesel engines, the Land Rover 2.25-litre four-cylinder diesel stood no chance at all, and Jaguar Rover Australia (as the local British Leyland subsidiary was then known) recognised that they were unable to compete in a potentially very lucrative sector of the market.

As a result they sought clearance from Solihull to develop their own more powerful diesel Land Rover and put it on the market in September 1981. The basic vehicle in the Australian conversion was a Stage I V8 Series III model exported from Solihull in kit form without an engine. The kits were assembled in Australia and JRA installed a Japanese Isuzu 4BD1 direct-injection diesel engine. This was a big overhead-valve iron-block four-cylinder of 3,856cc which had originally been designed to power a seven-tonne truck and it offered 96.5bhp at 3,200rpm and 187 lb/ft of torque at a very low 1,900rpm. It was a noisy and unrefined engine, but so successful was the conversion that by the time production of the 110 had started at Solihull in 1983, the Isuzu-powered 109 was accounting for some 70% of long-wheelbase Land Rover sales in Australia.

When the 110 was launched in Australia in November 1984, the only engines available were a high-compression (9.35:1) version of the petrol V8 and the Isuzu diesel, still rated at 96.5bhp and fitted with the LT95 four-speed gearbox. Payload of the diesel 110s was lower than that of the petrol V8s, because the Isuzu engine was considerably heavier than the Rover unit, but this did not seem to affect sales. During March 1985, JRA announced an additional model which could be had only with the diesel engine. This had a 120-inch wheelbase and was designed as a chassis/cab for utility bodywork to compete against the diesel Toyota utilities which were then dominating the mining and rural markets.

The additional model was built at JRA's Moorebank assembly plant in Sydney and was usually fitted with a tray-back body. The extra wheelbase length helped to reduce the rear overhang commonly associated with this type of body and therefore to improve the vehicle's rough-terrain ability. The 120-inch model benefitted from a galvanised chassis but lacked the rear anti-roll bar fitted to Australian-assembled 110 Station Wagons (and not found on Solihull-built vehicles). Payload, however, was only one tonne.

By the end of 1985, Australian 110 V8s had the five-speed LT85 gearbox, and this was also fitted to the diesel-powered models from January or February 1986. The 90 never was put on sale in Australia and the 110 County and 120 chassis/cab models in both V8 petrol and Isuzu diesel-engined forms remained the only Land Rovers available in that market until June 1987, when JRA announced what it called the 110 County pick-up or Land Rover Dual-Cab - a strange hybrid which consisted of a four-door cab constructed mainly from Station Wagon parts and a shortened pick-up back body. The vehicle came with either five or six seats and could be had with either the petrol V8 or the Isuzu diesel engine. However, few were made: the vehicle appears to have been designed originally for the Australian Bicentennial Authority, who took a batch, but the only dealer which sold examples to the public was Winterfaulls of Perth.

In the meantime, JRA had also designed and developed yet another unique Land Rover variant, a three-axle six-wheeler with six-wheel drive and a specially widened cab. The idea for a 6x6 Land Rover actually came out of some market studies done in the late 1970s which showed that there was a demand in Australia for a model with a higher payload than the then current 109 Series III or the 110 which JRA knew to be under development. However, the spur to turning this idea into reality came in 1982, when the Australian Army announced Project Perentie - the quest for a fleet of all-terrain vehicles to replace its existing Land Rovers. The requirement was for 2,600 one-tonne vehicles and for 400 two-tonners, and JRA, already planning to enter the 110 into the one-tonne trials, decided to develop a vehicle to meet the two-tonne requirement as well.

The new vehicle was developed entirely in Australia, although Land Rover in the UK funded the project, set up the test parameters, provided specialised design assistance and supplied special components. The project was led by Ray Habgood, JRA's Land Rover Engineering and Product Planning Manager, and the first prototypes (which used Stage I front end panels) were ready for the Australian Army trials which began in October 1983.

To create a two-tonne payload capability, JRA had extended the rear chassis and fitted a third axle; the distance between the first and second axles was the same 119.68

The Isuzu 3.9-litre diesel engine.

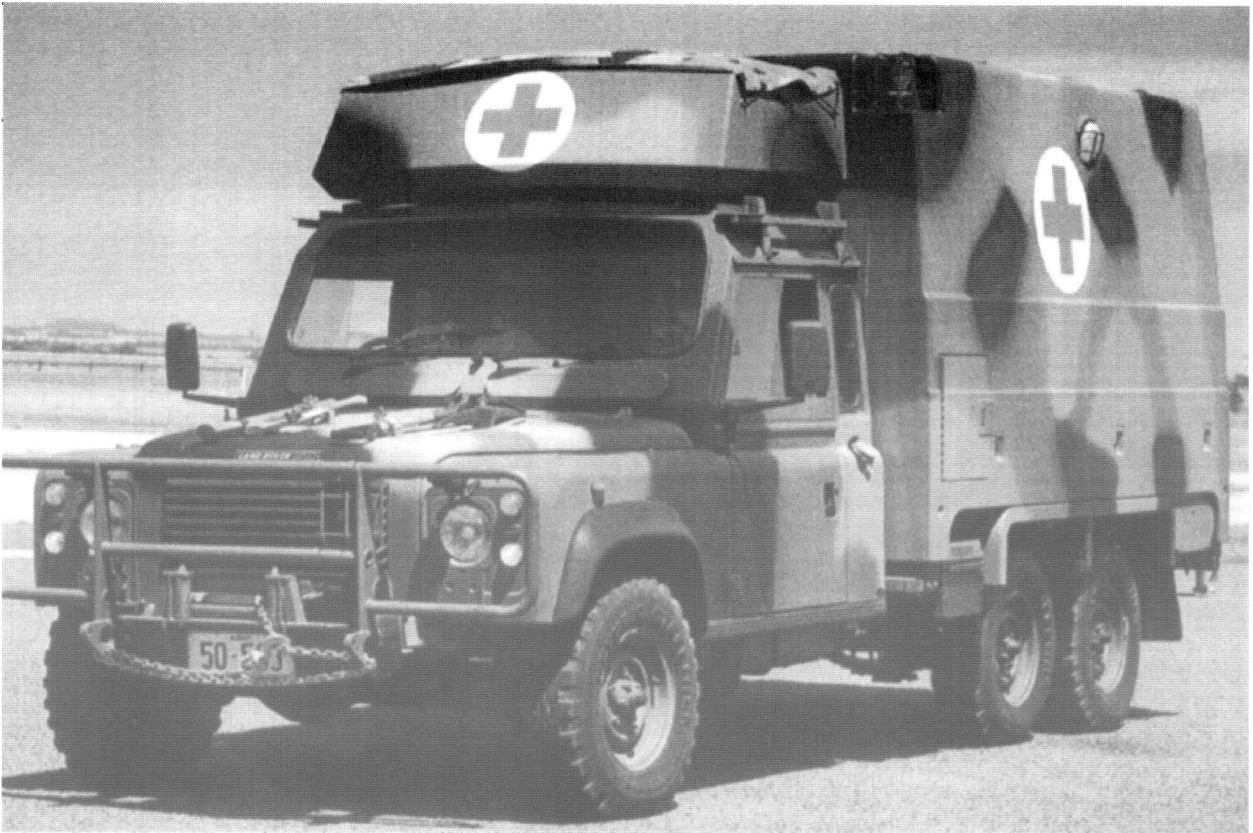

A hard-skin military version of the JRA 6x6 with the special wide cab.

This view of a JRA 6x6 under construction shows the unique construction of the rear chassis frame.

inches (3,040mm) as on the JRA "120-inch" model. However, the whole chassis was actually quite different from the 110s with deeper side members and cross members at the front and a fabricated section at the rear built up from square tube and channel section steel. After completion, the whole frame was hot-dip galvanised. Several different rear suspension layouts had been tried during development, but JRA finally settled on a twin-axle bogie arrangement, with semi-elliptic springs on each axle linked by a load-sharing rocker beam. This gave excellent axle articulation for off-road work together with good resistance to body roll on the road.

As the Australian Army wanted diesel power, JRA equipped their 6x6 with the Isuzu engine (although the petrol V8 was also offered when the vehicle was made available in other markets at the end of the 1980s). However, the 6x6 was equipped with the more powerful turbocharged version of the engine, which had 115bhp and 235 lb/ft of torque at 1,800rpm. An LT95A four-speed gearbox was fitted with permanent drive to the front and second axles, and drive to the third axle was automatically engaged when the centre differential was locked: the third propshaft was driven from the central power take-off on the transmission.

The front axle was the same as that of the 110, but strengthened by means of stiffening tubes pressed inside its housing, and a stronger four-pinion differential was fitted. Constant-rate coil springs were used in place of the 110's dual-rate type and lifted the ride height of the front end by 25mm (just under an inch). At the rear, both axles were wide-track versions of the 110's Salisbury 8 HA fully-floating type with strengthened casings and long-travel dampers developed specially for the 6x6 by Monroe-Wylie. Even the wheels were specially developed, looking like the standard items but actually being thicker in the nave area. In due course, JRA would offer the vehicle with options to suit either 15-inch or 16-inch tyre sizes and also with a locally-made split-rim option.

The 6x6 was designed to carry different body modules so that the basic vehicle could be converted in the field from 12-seat troop carrier to an ambulance, a water tanker or a fire tender. The canvas top, hood sticks, and longitundinal seats in the load bed could all be removed and stowed below the tray and an alternative body module could simply be lifted into place and bolted down.

In July 1986, the Australian Army announced that JRA had won the Perentie contract to supply vehicles in both the one-tonne and two-tonne categories. However, one condition was that the 6x6 should come with an enlarged cab - and in fact the prototype of that cab was under construction in JRA's workshops shortly after the announcement was made. The "wide cab" which went into production was essentially a standard Land Rover cab widened by 200mm (7.8 inches), lengthened by around 2.5 inches and with its roof raised by around 2 inches to give additional headroom. In addition, it had a deeper rear window to improve rearward visibility. The standard 110 facia panel was used and the extra space at either end was filled with stowage clips for rifles. A new bonnet panel also had to be designed and JRA made that up from GRP with a welded steel supporting structure.

Production of the 6x6, or "110 Heavy-Duty 6x6" as it was properly known, began in 1986 and JRA sold small numbers on the civilian market in Australia - most notably to mining companies as fire tenders and to tour companies as the basis of outback buses. From May 1989 it was also offered through Land Rover in Solihull to overseas military authorities.

The 110s which JRA supplied to the Australian Army were also special in their own way. All had the naturally-aspirated Isuzu diesel engine and all had certain modifications to meet the military's requirements. As the Australian Army did not want the spare wheel to be carried inside the body or on the bonnet, JRA splayed the rear chassis members to make room for it under the tail and relocated the fuel tank to compensate. Like the utility 110s built in Solihull, the Australian Army's vehicles had no self-levelling strut on their rear axles.

The final deliveries from JRA to the Australian Army under the original Perentie contract were expected to be made during 1994, by which time the original 110s had long been replaced on the civilian market by Defender 110s.

The extra propshaft running to the third axle is clearly visible in this picture of a JRA 6x6 chassis.

FACTORY-APPROVED CONVERSIONS

The tradition of factory-approved conversions was well established by the time the coil-sprung Land Rovers came on the scene in 1983. Beginning in the mid-1950s, specialist body builders and equipment manufacturers had seen in the Land Rover an ideal vehicle on which to base conversions, but they had been anxious to ensure that their conversions did not invalidate the factory's standard warranty. So it was that a system of "factory approval" for these conversions was drawn up in the late 1950s. When the 110 was introduced, the department which dealt with approved conversions was called Special Projects; from July 1985, however, it was reorganised as Special Vehicle Operations (SVO) and began also to design and build its own special bodies for a variety of purposes.

It would be impossible to cover the full range of factory-approved conversions in a book of this size, but the pictures which follow give a general idea of their variety and scope. The majority of specialist conversions were built on the 110 or 127 chassis because these offered maximum space.

Fire appliances have always given the specialist converters scope for building some interesting vehicles. This one was built in 1984 by Carmichael on one of the earliest Crew Cab 127-inch chassis. It has the V8 petrol engine.

A different approach to fire appliances is seen on this one built by Pilcher-Greene on a 110 chassis in the mid-1980s.

MMB International built this ambulance on a left-hand drive 110 chassis.

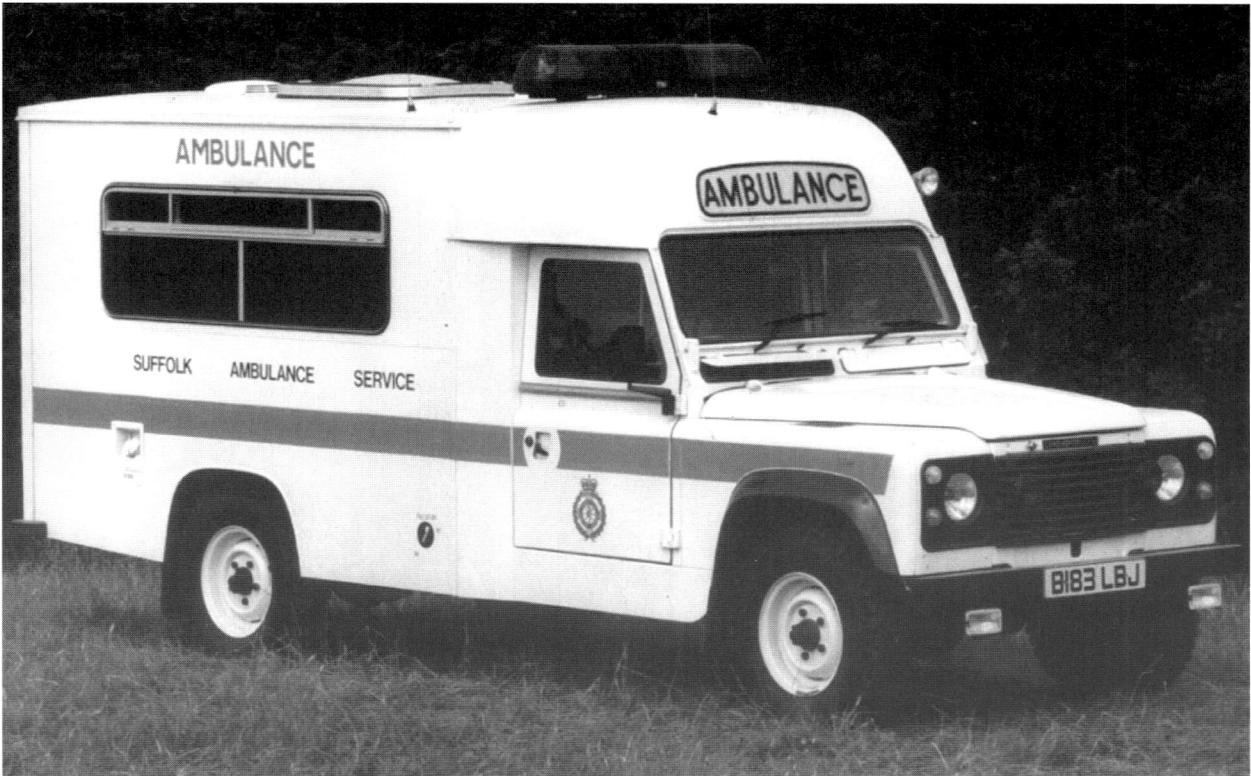

This ambulance was built in 1984 or 1985 for the Suffolk Ambulance Service on a 127-inch chassis.

Both the 90 Hard Top and 110 Station Wagon found favour with Police Forces. This Thames Valley Police 110 was equipped as an Accident Unit and carries a Dale Stem-Lite, which provides overhead floodlighting for an accident scene at night. It is based on a 1984-model Station Wagon.

This Devon and Cornwall Police Command and Communications Vehicle dates from 1986 and is based on a 127-inch Crew Cab chassis. The rear body is SVO's own high-roof box type.

SVO's box body came with high roof or low roof and to suit chassis/cab or crew cab configurations of the 127 chassis. This is a low-roof version, fitted to a crew-cab 127.

This 110 HCPU has been fitted with a hydraulic platform lift for overhead servicing work.

This armoured Cash Guard bullion carrier was based on a 110 Hard Top in the mid-1960s.

MODIFICATIONS

By the time the coil-sprung Land Rovers appeared on the market in the early 1980s, modifying Land Rovers was already big business. That business steadily increased during the 1980s, as the public began to see Land Rovers more as passenger-carrying vehicles which could be individualised and less as purely utility vehicles. Land Rover themselves set up a new subsidiary, Land Rover Parts and Equipment (later simply Land Rover Parts) to capitalise on this business and a vast array of accessories - some practical and some purely cosmetic - became available for the coil-sprung models.

This is not the place to list every single accessory which was ever available for the Land Rovers of the 1980s; besides such a list would rapidly become out of date as new accessories are made available for these vehicles in their later working lives. Fashions and customer requirements change, and accessories change with them. Back in the mid-1960s, for example, who would have thought that the then-current Series IIA models might be updated 20 years later with bullbars and side decals, neither of which could be had when the vehicles were new?

Nevertheless, the relatively simple construction of the coil-sprung Land Rovers, together with the wide interchangeability of their components, means that it is relatively easy to make major modifications as well as to bolt on cosmetic accessories. Thus, pick-up bodies can be swopped for Station Wagon types without too much trouble, thirsty V8 engines can be swopped for frugal diesels and four-speed gearboxes can be swopped for five-speeds. Magazines such as Land Rover Owner regularly run how-to features on such transformations, which of course are beyond the scope of the factory workshop manuals.

Many owners have carried out diesel engine transplants, commonly using any one of a number of non-original types which are more powerful or more refined than the 2.25-litre and 2.5-litre engines which the factory fitted. Engines by Perkins and Peugeot were often used in the early days, but more recently the well-developed high-speed types from Japanese manufacturers such as Mazda, Nissan and Isuzu have dominated the market. Land Rover Parts also made a strong bid in this market in 1993 by offering an attractively priced retro-fit kit for their highly-regarded Tdi engine, which of course was the standard diesel engine in the Defender models which

replaced the original 90,110 and 127 in 1990.

A few owners have also fitted tuned V8 petrol engines to improve the road performance of their vehicles. The tuning takes a variety of forms - larger-displacement blocks, fuel injection (from the Range Rover engine) in place of carburettors, turbochargers and even superchargers. Another favourite way of gaining better performance is to fit a Ford V6 engine (usually the early 3-litre type) and professionally engineered kits exist to make the conversion easy to do at home.

Transmission modifications are very much less common than engine modifications because the standard five-speed gearbox is well up to the demands made by most owners. However, some owners have wanted automatic transmissions and there are several vehicles around with the ZF four-speed automatic box from the Range Rover and Discovery. Generally speaking, automatics are fitted with the V8 petrol engine or in conjunction with a Tdi diesel transplant; the other standard engines are not powerful enough to suit an automatic box.

Body modifications do not usually depart radically from the standard factory offerings and are mostly cosmetic. However, seating and interior fittings have often been upgraded, usually in pursuit of greater comfort or a more luxurious ambience. Additional seats may have been fitted in van bodies, and forward-looking seats have also been available to replace the inward-facing type in the rear of Station Wagons.

It is worth briefly touching on the question of tyres, if only because it is a subject so regularly misunderstood. The tyres fitted as standard equipment to the coil-sprung Land Rovers were a dual-purpose type, designed to minimise noise at high road speeds and yet to have a sufficiently aggressive tread pattern to give good off-road traction. Some owners have fitted wider tyres with special wheels to suit, primarily to give their vehicles a more aggressive appearance. Such tyres will only rarely improve off-road traction and will probably make the on-road ride worse than standard, though they may improve the roadholding. Others fit special tyres (such as Firestone SATs) for improved off-road ability. However, there is no point in making such modifications unless the vehicle is going to be used primarily for heavy off-road work: the standard tyres are good enough for all but the most demanding types of terrain.

BUYING A CONVERTED OR MODIFIED VEHICLE

Very many coil-sprung Land Rovers have been modified or converted in one way or another and one of the purposes of this book is to help buyers to identify what is original equipment and what is not on a Land Rover offered for sale. The variations are infinite and it would be possible to do no more than scratch the surface of them here. However, a few guidelines may be helpful.

Converted vehicles - that is, those with major adaptations which enable them to perform specific tasks such as fire tenders, ambulances or breakdown vehicles - do not come up for sale very often. To the average Land Rover buyer, they are of no interest unless they can be stripped down to a bare chassis and built up again with more suitable bodywork. They can often be bought very cheaply and, when rebuilt with second-hand bodywork of the preferred type, can often prove cost-effective purchases. However, they sometimes incorporate hidden modifications which may prove troublesome to reverse, and so it is advisable to be very cautious before buying such a vehicle.

Former military vehicles may not have been converted in such obvious ways, but buyers need to be aware that there was a special military specification and that not everything on an ex-military Land Rover will be the same as on civilian versions. In particular, many military Land Rovers were equipped with 24-volt electrical systems as "FFR" (Fitted for Radio) vehicles. These have unique electrical systems for which many parts are prohibitively expensive and are best avoided unless the intention is to convert them to standard 12-volt specification.

Modifications, which by their very nature tend to be less radical than conversions, will be less obvious. In all cases the first thing to do is to establish exactly what has been modified. The next thing a buyer should do is to decided whether he or she actually wants to retain that modification. If not, is it reversible? Modifications which are likely to leave scars if reversed can prove useful bargaining points when haggling over the price of a used Land Rover.

If a modification is to be retained after purchase it is important to establish exactly what has been done and where replacement items can be obtained in case of breakage. Electrical modifications are notorious for giving trouble and it may take a very long time to trace additional wiring when a fault develops. Other add-ons need to be examined carefully with an eye to the quality of their manufacture because, although most of the aftermarket suppliers have produced well-made accessories for the Land Rover in the certain knowledge that they are likely to be treated roughly, there have been some which have proved rather flimsy in use. It is also true that some of the smaller aftermarket manufacturers have not survived very long in business with the result that it is no longer possible to buy replacements for items which they sold.

Finally, it is worth remembering that some modifications - particularly those which increase the acceleration or maximum speed of a vehicle - may result in higher insurance premiums than for a standard vehicle. Many insurers insist that owners declare all changes from standard specification on a Modified Vehicle Report form and some will ask for an Engineers' Report as well. In all cases, it is advisable to check what the insurance requirements are because an insurance company can refuse to meet claims if modifications which materially affect the risk have not been declared.

90s AND 110s IN
THE CAMEL TROPHY

A pair of 90s wade through a river during the 1986 Camel Trophy in Australia.

Thoughout the period when the 90 and 110 were current production vehicles, Land Rover promoted an image of their ruggedness and individualism by linking them closely with the annual Camel Trophy adventure rally. Although Range Rovers were used on the 1983 and 1987 events and Discoverys on the 1990 event, diesel Station Wagon versions of the Land Rover utilities were used on all the others. In 1984, the Camel Trophy crews used 110 diesels in 1985, and 1986 the vehicles were 90 diesels, and in 1988 and 1989 the crews were provided with 110 Diesel Turbos.

In essence, the Camel Trophy was (and remains) a 1,000 mile convoy rally through some of the world's most inhospitable terrain. Typical were the jungles of the Amazon Basin (where the 1984 and 1989 events were held), or the outback in Queensland, Australia (which played host to the 1986 event). Each vehicle carries a crew of two from one of the participating nations, and each crew is selected after a series of gruelling trials. The qualities need in a crew member range from physical stamina through intelligence to navigational and driving ability, but the ability to work as part of a team is paramount. And in fact the competitive element of the Camel Trophy is very much subordinate to the need to get the whole convoy through difficulties encountered en route.

In addition to the actual team vehicles, Land Rover has always supplied a quantity of support and back-up vehicles which have carried supplies, spares and medical aid - all of which have often been used to assist the local inhabitants as well as the crews participating in the event. All the vehicles have been painted in the distinctive sand yellow colour of the event's primary sponsors (R.J.Reynolds Tobacco, makers of Camel cigarettes) and have worn appropriate decals. All the team vehicles have been kitted out identically with equipment such as extra lighting, sand ladders, roll cages and so on, but the support vehicles have of course been differently equipped to suit their individual roles.

Some former Camel Trophy vehicles have been sold off after the events; others have been donated to various local authorities or projects in the event's host countries. The event's image and appeal have been powerful enough to persuade some owners to create Camel Trophy lookalike vehicles and there remains a risk that some of these may one day be passed off as the real thing. When buying a vehicle which is described as ex-Camel Trophy, it is important to remember that it could be one of three types:

   i) a genuine ex-Camel Trophy team vehicle
   ii) a genuine ex-Camel Trophy support vehicle or
   iii) a fake.

In cases of doubt, the sensible thing to do is to make a note of the vehicle's details (registration number and VIN) and to check with either Land Rover Ltd (021 700 2424) or with Global Event Management (0483 775500).

Fully equipped for the 1988 event in Sulawesi is this 110 Diesel Turbo. Note the sand ladders bolted to the roof rack, the jerry-cans for extra fuel and the additional lighting. There is a "snorkel" breather for the engine together with a brushguard, a Warn winch, and tools for digging when the going gets tough.

The last Camel Trophy to use 110s as crew vehicles before the Discoverys took over was the 1989 event in the Brazilian Amazon. This 110 returned to the UK, was refurbished as necessary, and was then sold off.

KEY DATES

Land Rover's habit of introducing major specification changes at irregular intervals makes the overall development of the 90 and 110 ranges rather difficult to understand. This list of key dates in the production history makes the sequence of changes easier to grasp:

March	1983	110 introduced
January	1984	2.5-litre diesel engine replaces 2.25-litre type
June	1984	90 introduced; 110 facelifted
May	1985	90 V8 introduced
August	1985	2.5-litre petrol engine replaces 2.25-litre type
October	1986	Diesel Turbo and uprated V8 petrol engines introduced
December	1987	whole range facelifted
December	1988	ribless roof and rivet-free upper body sides introduced
September	1990	range replaced by Defender 90,110 and 130 derivatives

Land Rover's new Land Rover

MISCELLANY

The Cariba concept vehicle was built in 1987

During 1983, Land Rover developed a 6x6 version of the 110 which used Range Rover axles at the rear. This was intended primarily for military use overseas but was also made available through Special Vehicle Operations for civilian use. It remained quite rare. The six-wheel-drive system was based on that developed by SMC for their Sandringham Six conversion of the Series III 109-inch Land Rovers.

In June 1986, Land Rover announced that a new Forward Control model based on the 110 chassis and drivetrain was undergoing trials with the British Army as a potential replacement for the 101 1-tonne Land Rover. However, the Army preferred the Reynolds-Boughton RB44 which was also competing for the contract and Land Rover decided not to go ahead with production of their new vehicle. In all, just ten prototypes and one line-built production vehicle were made. The new Forward Control, which had a forward-hinged cab made of GRP on a metal frame, was codenamed Llama. Several examples still exist in museum collections.

In autumn 1987, Land Rover built a special one-off "sunshine concept car" based on a V8 90 and called the Cariba. This was primarily done as a PR exercise and although public and press reaction was extremely favourable the company did not put it into production. The reasons, probably, were that it would be too expensive and that resources were already fully committed to the forthcoming Discovery. However, the idea was not lost: the limited-edition Defender SV90 which appeared in 1992 was very much along the same lines as the Cariba concept vehicle.

When Land Rover announced that they were not planning to put the Cariba into production, several specialists looked at the feasibility of producing similar vehicles to meet demand. Tractamotors of Melton Mowbray announced their TRM 350 C in the early summer of 1988, and this was essentially a soft-top 90 with improved trim and paint and a Stage I tuned V8 engine. At about the same time, the Land Rover Centre in Huddersfield announced a Cariba lookalike based on a 90 but fitted with a 2.25-litre diesel or a Ford V6 engine. Bearmach of Cardiff, the aftermarket parts specialists, also built a demonstrator which transferred the Cariba concept to a Series III 88-inch model. None of these sold in any quantity.

Land Rover was always looking for overseas military contracts, and between the mid-1970s and about 1988, it conducted protracted negotiations with the French and Swiss military authorities in the hope of securing a contract. A whole series of trials vehicles was built during this period, all with 100-inch coil-sprung chassis; some had automatic transmissions with V8 petrol engines. However, no contract was ever signed and the 100-inch military Land Rovers never did go into production. Several still survive, both in museum collections and in private hands.

In 1989, the London coachbuilders Vantagefield constructed two luxury models based on the 110 Station Wagon. These incorporated front and rear styling changes which used Range Rover body parts and had leather-trimmed interiors with Range Rover seats.

Tractamotors' TRM 350 C was an attempt to emulate and develop the concept behind the Cariba.

The Llama Forward Control prototypes had 110-inch wheelbase chassis with coil springs and used drivetrain components from the 110.

A 100-inch military trials vehicle with automatic transmission which was built in the hopes of securing a contract with the Swiss Army.

One of the two Vantegfield Coyote models built in 1989

The rare One Ten 6x6 is seen here equipped with a Spencer 8-28 Articulated Boom Platform.

LAND ROVER VATBACK.

The One Ten County is the ultimate people carrier.

12 seats . . . powerful V8 engine, or 2.5 litre petrol and diesel options . . . and refined suspension for supreme comfort.

All this with Land Rover's pedigree and legendary 4 wheel drive ability.

Surprisingly, though, the One Ten County still qualifies for mini bus status. So there's no car tax, and you can claim the V.A.T. back. Leaving a price tag no bigger than an upmarket estate.

If you want to carry more, with less overheads, test drive the remarkable One Ten Vatback.

SPECIFICATIONS
HOME MARKET MODELS

110 models

March 1983 to December 1984

Engines: 2,286cc (90.47mm bore x 88.9mm stroke) OHV four-cylinder petrol, with 8:1 compression ratio and Weber 32/24 DMTL twin-choke carburettor. 74bhp at 4,000rpm and 163 lb/ft at 2,000rpm.

2,286cc (90.47mm bore x 88.9mm stroke) OHV four-cylinder diesel, with 23:1 compression ratio and CAV injection pump. 67bhp at 4,000rpm and 103 lb/ft at 1,800rpm.

3,528cc (88.9mm bore x 71.1mm stroke) OHV vee-eight petrol, with 8.13:1 compression ratio and two Zenith-Stromberg carburettors. 114bhp at 4,000rpm and 185 lb/ft at 2,500rpm.

Transmission: Four-cylinder models have LT77 five-speed main gearbox with LT230 two-speed transfer box; eight-cylinder models have LT95 four-speed main gearbox with integral two-speed transfer box. Permanent four-wheel drive with lockable centre differential standard on all models; selectable two-wheel (rear) drive optional on four-cylinder models.

Gear ratios (five-speed): 3.585:1, 2.30:1, 1.507:1, 1:1, 0.83:1, reverse 3.701:1; transfer gears 1.667:1 (High range) and 3.320:1 (Low range).

Gear ratios (four-speed): 4.069:1, 2.448:1, 1.505:1, 1:1, reverse 3.664:1; transfer gears 1.336:1 (High range) and 3.321:1 (Low range).

Front and rear axle ratios: 3.54:1

Steering, suspension and brakes: Recirculating-ball, worm-and-nut steering with 20.55:1 ratio standard; power-assisted worm-and-roller steering with 17.5:1 ratio optional. Live axles front and rear with coil springs, dual-rate at the front, and hydraulic telescopic dampers; front axle located by radius arms and Panhard rod; rear axle located by radius arms, support rods and central wishbone assembly. Boge Hydromat self-energising rear ride-levelling strut standard on County Station Wagons and optional on other models. Dual-circuit hydraulic servo-assisted brakes, with 11.8 inch discs at the front and 11 inch drums at the rear; internal expanding drum-type parking brake operating on transfer box rear output shaft. 16-inch wheels with 7.50 x 16 crossply tyres or (County models) 205 x 16 radial-ply tyres.

January 1984 to May 1984

Engines: 2,286cc diesel replaced by: 2,494cc (90.47mm bore x 97mm stroke) OHV four-cylinder diesel with 21:1 compression ratio and DPS injection pump. 67bhp at 4,000rpm and 114 lb/ft at 1,800rpm.

June 1984 to April 1985

Transmission: Selectable two-wheel drive option for four-cylinder models no longer available.

May 1985 to July 1985

Transmission: Five-speed LT85 main gearbox with LT 230 transfer box replaced LT95 four-speed gearbox with integral transfer box on V8 petrol models.

Gear ratios: 3.65:1, 2.18:1, 1.43:1, 1.1:1, 0.79:1, reverse 3.82:1; transfer gears 1.41:1 (High range) and 3.32:1 (Low range).

August 1985 to September 1986

Engines: 2,286cc petrol replaced by: 2,494cc (90.47mm bore x 97mm stroke) OHV four-cylinder petrol with 8:1 compression ratio and Weber twin-choke carburettor. 83bhp at 4,000rpm and 133 lb/ft at 2,000rpm.

October 1986 to August 1990

Engines: 3,528cc V8 petrol engine now with two SU carburettors, 134bhp at 5,000rpm and 187 lb/ft at 2,500rpm.

Existing diesel engine supplemented by: 2,494cc (90.47mm bore x 97mm stroke) OHV four-cylinder diesel with 21:1 compression ratio, DPS fuel pump and Garrett AiResearch T2 turbocharger. 85bhp at 4,000rpm and 150 lb/ft at 1,800rpm.

90 models

June 1984 to April 1985

Engines: 2,286cc four-cylinder petrol and 2,494cc four-cylinder diesel; specifications as for contemporary 110.

Transmission: Five-speed LT77 main gearbox with two-speed LT230 transfer gearbox; specifications as for contemporary 110 except 1.41:1 High ratio in transfer gearbox. All models had permanent four-wheel drive.

May 1985 to July 1985

Engines: Four-cylinder petrol and diesel types supplemented by 3,528cc V8 petrol engine; specifications as for contemporary 110.

Transmission: 3,528cc V8 petrol models had LT85 five-speed main gearbox with two-speed LT230 transfer gearbox. Specifications as for contemporary 110.

August 1985 to September 1986

Engines: 2,286cc petrol engine replaced by 2,494cc petrol engine; specifications as for contemporary 110.

October 1986 to August 1990

Engines: 3,528cc V8 petrol engine now uprated to 134bhp, as for contemporary 110. 2,494cc Diesel Turbo added as fourth engine option; specifications as for contemporary 110.

Crew Cab and 127 models

Specifications generally as for contemporary 110 models.

V E H I C L E   I D E N T I F I C A T I O N

All Land Rover 90,110 and 127 models have their chassis number stamped on a plate fixed to the bonnet shut panel. A typical chassis number for one of these vehicles would be SALLDHBV2AA101001.

These 17-digit chassis numbers consist of an 11-digit prefix or VIN (Vehicle Identification Number) code, followed by a six-digit number which is the actual serial number of the vehicle. Serial numbers began at 100001 for Solihull-built vehicles and 500001 for CKD types, but in May 1986 the 90 and 110 models were incorporated into the same number sequence as Range Rovers (and Discoverys were added to this sequence just over three years later). The new sequence began at 261902.

The VIN codes remained unchanged in May 1986 and conform to a standard pattern now in use by manufacturers all over the world. They reveal a number of important details about the vehicle's build specification. The VIN code does not reflect any modifications which may have been carried out after the vehicle left its manufacturer's premises.

The VIN codes break down as follows:

SAL: Land Rover Ltd (manufacturer's identity code)
LD: 90 or 110 (model code)
H (etc) : Wheelbase length. The codes used are H (110 inches), M (Special), V (92.9 inches, or "90") K (127 inches)
A (etc): Body type. The codes used are A (basic models), B (two-door Station Wagon), H (High Capacity Pick-Up) and M (four-door Station Wagon)
C (etc): Engine type. The codes used are C (2.5-litre diesel), D (2.5-litre petrol), G (2.25-litre diesel), H (2.25-litre petrol) and V (3.5-litre petrol)
1 (etc): Steering and transmission type. The codes used are 1 (RHD, four-speed), 2 (LHD, four-speed) , 7 (RHD, five-speed), and 8 (LHD five-speed)
A (etc) : Model type. The codes used are A (1983-84), B (1985-1987 model-years), E (1988 model year), F (1989 model year), and G (1990 model year)
A (etc): Assembly location. The codes used are A (Solihull) and F (overseas assembly from CKD set)

Although it is not possible to determine the build-date of a vehicle very accurately by means of its chassis number, the following serial numbers provided by Land Rover may be helpful:

184732	March 1983	First 110
208595	January 1984	
213333	June 1984	First 90
229956	January 1985	90
230001	January 1985	110
241937	August 1985	First 110 with 2.5-litre petrol engine
256479	January 1986	110
256593	January 1986	90
281194	January 1987	
314041	January 1988	90
314369	January 1988	110
358393	January 1989	110
358791	January 1989	90
421115	January 1990	110
423950	January 1990	90

Engine numbers will be found on the right-hand side of the cylinder block (four-cylinder types) or on a ledge beside the dipstick on the left-hand side of the cylinder block (V8 types). The type identifiers are as follows:

10H	2.25-litre petrol, 8:1 compression, non-detoxed
10J	2.25-litre diesel
11H	2.25-litre petrol, 8:1 compression, detoxed to ECE 15-03
12J	2.5-litre diesel
13H	2.25-litre petrol, 7:1 compression
14G	3.5-litre petrol, 8.13:1 compression, non-detoxed
15G	3.5-litre, 8.13:1 compression, detoxed
17H	2.5-litre petrol, 8.13:1 compression, detoxed
19G	3.5-litre petrol, 8.13:1 compression, detoxed to Saudi Arabian specification
19J	2.5-litre diesel turbo
20G	3.5-litre petrol, 8.13:1 compression, non-detoxed
21G	3.5-litre petrol, 8.13:1 compression, detoxed
22G	3.5-litre petrol, 9.35:1 compression, Australian specification
24G	3.5-litre petrol, 8.13:1 compression, detoxed

Gearbox numbers will be found stamped on the casing near the drain plug. Identifying numbers are:

13C	four-speed LT95 (early V8 only)
20C	five-speed LT85 "Santana"; early type
22C	five-speed LT85 "Santana"; late type, lightweight with divided casing
50A	five-speed LT77 for 2.25-litre petrol and diesel engines

Transfer boxes were stamped with serial numbers on the casing near the filler plug. Identifying numbers are as follows:

10D	selectable four-wheel-drive (early 2.25-litre petrol and diesel models only)
12D	permanent four-wheel-drive (2.25-litre petrol and diesel engines)
20D	permanent four-wheel-drive (2.5-litre petrol and diesel engines)
22D	permanent four-wheel-drive (2.5-litre petrol, diesel and diesel turbo engines)
25D	permanent four-wheel-drive (3.5-litre petrol engines, One Ten)
29D	permanent four-wheel-drive (3.5-litre petrol engines, Ninety)

Finally, axles also bear an identifying number, stamped on the tail of the differential housing (front axles) or alongside the rear of the differential housing (on rear axles). These numbers are:

20L	front axle, right-hand-drive, One Ten
21L	front axle, right-hand-drive, Ninety
	front axle, right-hand-drive, One Ten
21S	rear axle, One Ten
22L	front axle, left-hand-drive, Ninety
22S	rear axle, Ninety
23S	rear axle, heavy-duty, Ninety

COLOUR CHART

Note: The colours given here are those offered as standard on the production lines. Many Land Rovers ordered for business use were of course painted to special order.

March 1983 to May 1984

Bronze Green	Roan Brown
Light Green	Russet Brown
Limestone	Sand
Marine Blue	Stratos Blue
Masai Red	Trident Green
Mid Grey	

All wheel centres and all fixed cab roofs, hardtop roofs and Station Wagon roofs were finished in Ivory.
All wheelarch eyebrows were finished in body colour.
All soft tops and tilts were finished in Khaki.
County stripes were always Cream.
All bumpers had a galvanised finish.

June 1984 to April 1985

Bronze Green	Roan Brown
Ivory White	Slate Grey
Marine Blue	Stratos Blue
Masai Red	Trident Green

All wheel centres and all fixed cab roofs, hardtop roofs and Station Wagon roofs were finished in Ivory.
All wheelarch eyebrows were finished in body colour.
All soft tops and tilts were finished in Matt Stone.
County stripes were finished in various colours.
All bumpers had a galvanised finish.

May 1985 to September 1986

Arizona Tan	Slate Grey
Bronze Green	Stratos Blue
Ivory White	Trident Green
Marine Blue	Venetian Red

All wheel centres and all fixed cab roofs, hardtop roofs and Station Wagon roofs were finished in Ivory.
All wheelarch eyebrows were finished in body colour.
All soft tops and tilts were finished in Matt Stone.
County stripes were finished in various colours.
All bumpers had a galvanised finish.

October 1986 to November 1987

Bronze Green	Stratos Blue
Ivory White	Trident Green
Marine Blue	Venetian Red
Slate Grey	

All wheel centres and all fixed cab roofs, hardtop roofs and Station Wagon roofs were finished in Ivory.
All wheelarch eyebrows were finished in body colour.
All soft tops and tilts were finished in Matt Stone.
County stripes were finished in various colours.
All bumpers had a galvanised finish.

December 1987 to November 1988

Arrow Red	Shire Blue
Arizona Tan	Slate Grey
Bronze Green	Trident Green
Ivory White	

All wheel centres and all fixed cab roofs, hardtop roofs and Station Wagon roofs were finished in Ivory.
Wheelarch eyebrows, grille panels and headlamp surrounds were finished in black on utility vehicles and in body colour on Station Wagons.
All soft tops and tilts were finished in Matt Stone.
County stripes were finished in various colours.
All bumpers were finished in black.

December 1988 to September 1990

Arran Beige	Ivory White
Arrow Red	Shire Blue
Arizona Tan	Trident Green
Eastnor Green	

All wheel centres and all fixed cab roofs, hardtop roofs and Station Wagon roofs were finished in Ivory.
Wheelarch eyebrows, grille panels and headlamp surrounds were finished in black on utility vehicles and in body colour on Station Wagons.
All soft tops and tilts were finished in Matt Stone.
Side decals on utility models were always in grey and black.
County stripes were finished in various colours.
All bumpers were finished in black.

ROAD PERFORMANCE FIGURES

These figures are averages, calculated from contemporary road tests and from figures claimed by Land Rover Ltd. They should be taken as a rough guide only, and do not of course reveal anything about the off-road ability of the 90 and 110 family of Land Rovers. Generally speaking, utility versions of all models are lighter than the County variants with the result that their acceleration and fuel economy can be slightly better.

Type	Max. Speed	0-60mph	30-50mph	Average mpg
90 County, 2.5 diesel, 67bhp	68mph			22
90 County, V8 petrol, 114bhp	85mph	14.5 sec	10.0 sec	15
90 County, V8 petrol, 134bhp	90mph	13.5 sec	10.7 sec	13
90 County, Diesel Turbo, 85bhp	75mph	22.5 sec	10.7 sec	18
110 County, V8 petrol, 114bhp, 4-spd	78mph	16.5 sec	11.0 sec	14
110 County, V8 petrol, 134bhp	85mph	15.1 sec	11.0 sec	13
110 Heavy-Duty 6x6, 3.9 turbodiesel, 115bhp	N/A	N/A	N/A	16

PRODUCTION FIGURES

The figures given here were kindly provided by Land Rover Ltd. They represent production totals and will therefore differ slightly from sales totals which have been published elsewhere. In each case, figures relate to calendar year and not to model-year. Figures for 1983-1985 will therefore include small numbers of the outgoing Series III models, while figures for 1990 include the first Defenders.

1983	28,412
1984	25,663
1985	23,772
1986	19,195
1987	20,475
1988	22,229
1989	22,738
1990	21,363